Cryptowriter

Volume One

Oct. – Dec. 2020

Published by The Writer Company Pty Ltd
writer.io
voice.com/cryptowriter
Cover Art by Lars Kommienezuspadt
www.darkpinup.com

All content published in this book was produced for Cryptowriter in association with Voice.

ISBN: 978-0-6450954-1-8

Table of Contents

Foreword

Written by Sean Ballent

The system is breaking. Journalists and local news are going extinct. The media is choosing sides, turning up the volume, and accelerating from the truth in opposite directions. Many aren't equipped to filter through the avalanche of noise.

Over the last decade, Facebook, Google, YouTube, and Twitter have made us their junkies. They've used us. Harvesting vast amounts of our data, we've become puppets of their narrative. And without better options, we just keep coming back for that next hit. The rip currents of algorithms and misinformation are pulling us out past the breakers, into uneasy and uncharted depths.

Blockchain is the answer. It's really that simple. Blockchain has the potential to hit the reset button and level the playing field. A disruptive force that will bring transparency and accountability to our world. Blockchain has the power to bank the unbanked, lift people, and align societies.

Sadly, this isn't the "blockchain" the media projects. Instead, they focus on clickbait, alarmism, and vanity metrics, with sensationalized headlines like "Exec Predicts $318,000 Bitcoin by Next December" and "DeFi Crash 2020: Is it really the end?" By selling out and playing with our emotions, the major crypto media outlets do their part to feed a bubble of uncertainty, driving much of the stigma that surrounds the industry. This form of sleazy coverage presents a considerable hindrance to blockchain adoption.

When Kenny Lienhard and I set off to create the crypto publication Cryptowriter, we wanted to deliver something genuine to the space. Our vision was to leverage the power of community and blockchain to deliver meaningful content that could be heard above the fruitless noise. We found a home for our publication on Voice.com, a blockchain-based social media platform where all users are actual people. At the center of Voice is the Voice token, which directly incentivizes content creators, like those in the Cryptowriter community to publish great content.

In order to get Cryptowriter off the ground, we went out and recruited a diverse team of crypto community experts on the basis that they would be empowered with the freedom to write about the things they're passionate about. We were surprised and humbled by the quality of people willing to take a leap of faith and join us on our mission. The result, a vibrant set of perspectives that are continuously exploring the ways blockchain is having a real impact on society and throughout our world. Over the last few months, I've learned a valuable lesson. Community matters.

Communities and blockchain matter in similar ways. They are reliable and can bring a sense of order to things when the surrounding world becomes disjointed. Let's face it, we humans are in a bit of a rut. We spend 7 hours a day staring at screens, flipping through feeds in search of that hit of digital dopamine. We've never been more connected, in more ways, but at the same time more divided. It's about the quality of those connections. When one becomes dependent on algorithms to dictate what they consume, they acquire a convoluted sense of reality and purpose. Communities provide quality connections and a sense of being to those within. Blockchains are trustworthy and keep us grounded.

The future is always uncertain, but blockchain's isn't. The revolutionary tech has taken root and will help us to cut through the noise and realign the world we live in. But like any disruptive technology, blockchain is only as powerful as the communities, people, and perspectives that drive it.

Introduction

Cryptowriter Volume 1 is a snapshot in time of 50 articles published in our Cryptowriter community between October and December 2020. Inside you'll find a diverse range of perspectives and opinions from cryptocurrency and blockchain experts. Join our crypto writers as they take you on a journey to explore the leading edge of a new frontier and what it means for you and the world we live in.

Cryptowriter lives on Voice.com, a new blockchain-based social media platform where every user is a verified human.

In order to give readers an opportunity to engage with the content in Cryptowriter Volume 1, we've provided QR codes at the beginning of each article. These QR codes are linked to the original source of their respective articles, on Voice.com. After registering for Voice, you can view comments and join the conversation by scanning any of the QR codes with your mobile device.

Gain full access to Voice by using our exclusive Cryptowriter invitation page via the QR code below.

Visit our community on voice.com via the QR code below.

Blox Populi: Blockchain – A Community Empowerment Toolkit

1

Written by Brandon Lovejoy

There's a shroud of mystery over the revolution of blockchain technology preventing everyday people from being aware of, or truly appreciating its potential.

Of course when one is saddled with debt, trying to keep the family in food, shuttling the kids to school, working long hours for little pay, helping out aging parents, enthralled to the 24-hour news cycle filled with shocking calls for attention, replete with scandals and elections, now nearly synonymous; well, it's hard to expect anyone to care about the revolution of blockchain technology.

Whenever a tool or technology comes along that threatens to put power in the hands of the people you can expect the ruling elites and the monolithic institutions that they control to create a sort of working global consensus to provide distractions, regulations, misinformation, anything to keep people from abandoning the systems which have kept them enthralled to the giant shell-game of global capital which has served the elites so well, for so long.

When Bitcoin burst into public consciousness, it was met with such widespread incredulity, hostility, and disparagement that we have still not fully recovered from the misperception that it is primarily a tool for criminals and various shadowy elements. And while Bitcoin has become the poster child of 'cryptocurrency', to this day when I mention the word blockchain to a new acquaintance, it is rarely met with any form of recognition. The underlying technology of Bitcoin has thus far been relegated to the world of tech gobbledygook.

Maybe we should be calling blockchain *'the community empowerment toolkit'*, or something a little more suggestive of its potential?

In an era when ordinary people have access to more powerful communication tools than ever before, such as the ability to easily and at a relatively low cost self-publish written, audio, or video content, and read the same thereby bypassing the well-established media conglomerates, the amount of noise is invariably amplified, lest people become well informed and more understanding as differing perspectives begin to creep in. Let's use blockchain to break through the noise!

While people have greater access to more new markets than ever before, they are largely driven like cattle to access just a few of the larger markets that leverage economies of scale to slake the modern thirst, and in many cases they need, for the most inexpensive and convenient options. Let's use blockchain to create markets that empower all participants!

The implementation of blockchain technology threatens to usurp several roles that the gatekeepers of society have long possessed. Blockchain allows network participants to freely transact between peers and renders the need for "trusted third-parties" superfluous.

The almost inevitable result of blockchain technology if applied to the social domain is enhanced transparency and integrity in any form of interaction, ie.

- Secure elections
- Verifiable news reported on by known entities with auditable reputations and disclosed conflicts of interest
- Customizable and adaptable monetary policies created by communities with fully auditable ledgers backed by a consensus of signed agreements between people
- Efficient access to medical records by medical professionals with proven and verifiable credentials
- Supply-chain management transparency and easy oversight of resource use and allocation
- The list goes on and on.

Blockchain really is a nearly magical community empowerment toolkit!

But not if we fail to deploy it.

Essentially, what blockchain can provide humanity if it will rise up and grasp it, is the basis for a civilization that dispenses with complicated bureaucracy and countless unnecessary obstacles to resolving the various challenges we face.

What blockchain demands of us, are the tools to facilitate easy access to experiment with and develop a new society built on the kind of unshakable integrity blockchain provides.

Governments and corporations of the world are scrambling like hell to retain their positions as arbiters of world commerce, macroeconomic decisions about resource use, ownership, movement, communications, and rights.

Communities who are now struggling with the loss of access to resources, habitat, sovereignty and a variety of natural 'rights' under the existing story of economic conquest will only benefit if they can effectively coordinate among each other to restore the role of managing resources and reaching agreements to the domain of the people.

"History, Stephen said, is a nightmare from which I am trying to awake." — James Joyce, Ulysses

Let's wake up humanity to the potential of new ways of being and doing.

Consider this another friendly reminder that the future is in our hands, so let's not pretend we never heard the good word that blockchain can be a tool to empower a global, community-driven, social transformation towards a sane and responsible monetary system, wise use of resources, clear agreements, and the healthy functioning of a truly free and open society not driven forward by antiquated competition among elites caught in the false notion of a winner take all struggle.

Let's put those short-sighted, self-serving interests out of business, let's do it together.

Peace, and @lovejoy

Africa - The Next Financial Powerhouse of the World?

2

Written by Jason Deane

How the African Continent is Poised to Take the Lead in the Use of Next-Generation Currency - and Why It May Not Get There.

We're totally spoiled in the West and we don't even realize it half the time.

Not only do we take basics like food, water, housing, education, and even electricity for granted, we also think of our banking facilities, as imperfect as they often are, as something that's just always there when we need it.

Want to order a pizza for delivery? No problem. In fact, you can use your saved credit card details to do it more quickly. Pay a bill online? Send money over to your mates for that weekend away? All easy and all, incidentally, only available in the last couple of decades, even though it's already completely entrenched in our modern lives.

In Africa, however, things are different. Only around 43% of the sub-Saharan population have access to any sort of electricity (source: Brookings), and only 34% have access to banking facilities (source: The World Bank).

In short, most people have neither. The figures are not much better for internet access, with only 39.3% having it, and seven of Africa's countries have access numbers of less than 10%. (source: internetworldstats.com)

Yet, even with these two apparent extremes juxtaposed against each other, there are many economic analysts and commentators, including myself, who see Africa as the "next big thing" in finance. How, then, can this be the case?

Africa—Land of Danger and Beauty

Africa, as a continent, is very much a caricature of human nature as a whole. It contains both the very best and the very worst of us, indifferently split across race and class, in a way that doesn't quite exist anywhere else.

Wealth, abject poverty, and everything in between is found within its borders, while beauty coexists with corruption, death, and a constant war for resources.

And those resources are abundant. Aside from having a substantial population, Africa is thought to hold around 30% of the world's known mineral reserves including cobalt, uranium, diamonds, and gold, as well as significant oil and gas reserves. By rights, Africa should be one of the richest nations on the planet.

But it isn't, and the reasons are largely due to political and historical problems that span centuries both within and outside the continent. There is little doubt, however, that Africa as a whole carries the potential to be a global power player, should that pathway ever be cleared.

And, for the first time, that just might be possible.

The Leapfrog

The concept of "leapfrogging" is well known among economists. The term is usually applied when a culture or entire country skips parts of a technical development process and jumps in at a later stage, thereby avoiding the growing pains associated with early adoption.

Those who successfully leapfrog usually benefit from a newer and more future-proofed system.

It has happened many times in many guises in many countries, but always in areas involving technology. In fact, it has already happened in Africa itself.

For example, Ethiopia was able to jump directly to hydroelectric power rather than going through the traditional coal-burning route first, and the continent as a whole was able to jump directly to mobile phone use without implementing landlines first.

Since the latter is already becoming obsolete in many Western countries, this jump stopped the nation from building a low-power hardwired network that is difficult to repurpose and allowed it to focus on an alternative, more modern system.

Africa even adopted the 3G standard before most Western countries did, as a result of that action.

Now, the same thing is happening with money and cryptocurrency, and it's entirely possible that Africa, whether by accident or design, is laying the groundwork for the biggest leapfrog in financial history.

The Mobile Money Boom

Africa, however, does not have a good track record with money. A seemingly endless supply of corrupt or incompetent governments has ensured that it is either repurposed for military or even personal use, spent on unnecessary projects, or simply printed to the point of hyperinflation.

Low banking penetration meant that many of its citizens were reliant on cash or barter for survival, and this was the way it was done for many years. In the last decade, however, things have changed.

It's so typical of Africa that there are parts of it where you can't get running water, but you can stream a movie on your mobile device. Despite the lack of consistent infrastructure, mobile phone penetration is surprisingly high with some countries—such as Nigeria and South Africa—having subscription levels as high as 9 in 10 people.

In other words, in these places, mobile devices are as common as they are in Europe or the U.S.

This created a classic leapfrogging scenario, where Africa's adoption of Mobile Money Operators (MMOs) was as fast as it was ubiquitous. These mobile services, supplied by operators such as

Ecocash, M-Pesa, OneMoney, Telecash, and MyCash, allow citizens direct access to electronic money.

For the first time, swathes of unbanked people could send money to each other by pressing a few buttons on their phones. It seemed to be a perfect solution.

However, recently, in true African style, Ecocash's services were suspended in Zimbabwe when it transpired that the system, owned by the monopolistic Econet Wireless company, was allegedly conducting a huge raft of illegal activities on a set of charges so long that writing them out would fill several sheets of A4 paper.

Since Ecocash handled more than 94% of the payment traffic, the impact of this would have been devastating to the ordinary citizen had it not been for the Reserve Bank of Zimbabwe (RBZs)'s intervention, making sure that peer-to-peer and merchant payments remained operational.

However, if you were a user of this service, you'd have to be concerned about what would happen next.

While you may have been completely happy with the speed, ease, and security of an electronic payment system over cash, especially if you have no access to traditional banking facilities, it's suddenly clear to everyone that you're still not in control of your money.

So, where do you go from here?

The Bitcoin Solution

Africa, generally speaking, has part of the problem solved already. Jumping straight to widespread mobile use over landlines and very high adoption rates of MMOs mean that the foundations of future monetary systems are already in place.

After all, this is unlikely to happen in a fully established culture where numerous alternatives coexist with a general apathy toward many of them. Why bother using a different service when the one you have works fine?

No, the problem is that this proliferation of use has occurred in a continent where there is unstable monetary policy, corrupt governments, and companies with far too much power and little regulation.

This means, for the ordinary man on the street, you have a service you can technically use quite easily, but you do so at your own risk. There are simply no guarantees, but also little alternative, so the demand still remains high despite this.

And, even if you do manage to stay in control of your money through the collective benevolence of the service providers, there's the constant pressure of devaluation, and the never-ending search for hard currency to preserve wealth.

What is required is a digital system that uses a hard, universally accepted currency, but also ones that lie outside of existing corrupt systems. What is needed is a system that gives peace of mind and total control back to those who use it, a system where confiscation or devaluation is out of the hands of those in power.

What is needed is Bitcoin.

And many parts of Africa have already realized it.

The Future is Bright?

Bitcoin use in Africa, as in the world generally, is increasing.

More than 14% of LocalBitcoins' and Paxful's global weekly trading volumes now come from Africa, with activity focused in Kenya, Nigeria, and South Africa. These volumes have seen significant growth this year, surpassing $10 million in weekly volume across the two platforms.

South Africa has arguably better pathways to Bitcoin than anywhere else across the country, and it is no coincidence that several exchanges and crypto orientated projects are based there. In fact, more have announced their intention to either open offices or increase their presence in the country.

The growing awareness of Bitcoin among the population is also borne out elsewhere. Google Trends data indicates that Uganda, Nigeria, South Africa, Kenya, and Ghana all rank in the top 10 on the topic of cryptocurrency.

South Africa actually ranked third-highest worldwide at 13%, with Nigeria ranking 5th (11%) in a survey about crypto ownership.

However, Coinshares data shows that only 0.2% of the world's Bitcoin nodes are located across the entire continent, and there is no meaningful mining activity at all.

For most Africans, therefore, their primary use case for Bitcoin is also Bitcoin's primary use case—that of a medium of exchange. Speculative or production activities such as supporting the network are clearly secondary or, may be more accurately described as "irrelevant" to most users.

The trends all seem to point perfectly to the obvious conclusion—that growth in Bitcoin use and awareness as a real, safe alternative to the tainted existing systems will continue, and Africa will find itself, by accident if not by design, at the forefront of the next financial revolution.

Unless, of course, it doesn't.

The Future May Not Be Bright

There are several main problems with this vision of the future, even in a continent as well placed as Africa, of being the first truly crypto-friendly location on earth.

Despite the proliferation of mobile technology, access to electricity and the internet remains sporadic and network improvements are coming very slowly. A recent UN report suggested that $100 billion would need to be spent over a decade to get national coverage to an acceptable level, money that is not currently available.

However, companies like SpaceX, Amazon, Viasat, and OneWeb are building low-orbit satellite mega-constellations that aim to provide high-speed internet across the globe, and companies like Blockstream and goTenna are pioneering Bitcoin transactions without internet via its mesh network.

These activities may be enough to fill the gaps in the short-term, especially if demand continues to grow.

But even if the infrastructure problems can be overcome sufficiently, there are still many questions on how Bitcoin adoption would unfold.

Governments face the same issues as their western counterparts—that is, how do you work with something that is not directly under your control? How do you tax it? How do you manage monetary policy?

If anything, these questions are far more relevant in an environment where many governments can't offer a solid alternative to their people, such as a stable fiat currency. The result? An increased propensity to be hostile to it.

In fact, as shown in the regulatory status map above, Libya, Algeria, Morocco, Zambia, Zimbabwe, and Namibia have already banned the use of cryptocurrencies, and while all other areas have no official stance as yet, several advise "caution" when using them. Only South Africa, Tunisia, Sierra Leone, and Senegal are actively positive in their approach.

It's feasible, then, that Africa's biggest obstacle to its own success in this regard could actually be its own regulatory structure. And if that's the case, that could be as devastating as it is frustrating for the continent.

Instead, let's hope Ban Ki-moon, the former UN secretary-general, was right when he said "the next century belongs to Africa"

Wall Street is About to Take Your Bitcoin

3

Written by Mark Helfman

I know a few people in finance. None of them fear cryptocurrency, even though the technology will make them obsolete.

Some plan to get out with what they can before cryptos take over.

Most want to hop on the train before it leaves the station. More and more with every passing day.

Does that get you excited? All those rich people and big investors bidding up the price of bitcoin. Feeding altseason. Spreading the wealth.

After all, this is the first time you have a chance to own a financial asset before "smart money" does. When else will you get a piece of the action before Wall Street does?

If you think "institutional investors" will send bitcoin's price to the moon, you're probably right. But at what cost?

You're talking about the savviest, best-connected, most deep-pocketed network of professionals on earth.

They're not getting into bitcoin because they care about open, permissionless, borderless, neutral, censorship-resistant money systems. They're getting into bitcoin to make money for themselves and their clients.

You're simply liquidity. You're the one who sells when they need to buy, buys when they need to sell, and pushes the value up long enough for them to cash out with a profit.

Dewey Cheatham and Howe, CFAs
In Bitcoin or Bust: Wall Street's Entry Into Cryptocurrency, I talk about all the ways they will use their money and power to take cryptocurrency away from you like they do with every other financial asset.

For example, they may tell your government to force everybody to use an "authorized entity" instead of private wallets. That authorized entity would have to meet certain requirements—requirements that only a Wall Street firm could meet. After all, we can't let bitcoin fall into the hands of criminals and terrorists, right?

Can you imagine new laws that force you to hire Fidelity as your "cryptocurrency service provider?" Wall Street can. Telecom did this to the internet in 1996. Why can't big finance do this to bitcoin now?

Or, maybe it will get your government to allow the commingling of cryptocurrency with other assets. While this lets Wall Street create new financial products—i.e., new ways to make money—it introduces massive risks for cryptocurrency markets.

As always, *you* take the risks while *they* get the money.

Maybe Wall Street will get the U.S. government to give it sole control over airdrops, ICOs, STOs, IEOs, and other activities that result in new cryptocurrency?

It's possible Wall Street firms will buy up all the small cryptocurrency exchanges and custodial wallet providers, integrate their wallets into new payment networks and effectively push light wallets out of the market. No more private keys.

And what happens when Bakkt sets up a node on the Lightning Network?

While Nobody's Watching . . .

Wall Street has a lot of money and power. Once it gets into the cryptocurrency markets, it will use this money and power for influence—marketing, lobbying, and networking.

Big firms will say "hey, remember 2017? The markets were awash with scams and frauds. We came in and cleaned the place up. It's safe now. Buy bitcoin (but buy it from us, not Coinbase, you can't trust them)."

They will try to get governments to regulate power and profits into their hands. After that, they will limit access, consolidate control, and build new risks into the financial system. It's the same playbook time and again.

Meanwhile, you and the rest of Crypto Twitter will continue to argue over central banks, monetary theory, and the role of government.

When you argue about fiat vs. bitcoin, Wall Street laughs. It's happy to make money on both.

<u>Six Conversations We Need to Have About Cryptocurrency Now But Won't</u>

It's Not the Bans You Need to Worry About

With so many ways for Wall Street firms to steer crypto into their hands, I'm amazed at how many people still worry about crypto bans.

Don't fear the bans. They never work.

Fear the lobbyists. They *always* work.

Somebody needs to create the tax laws, privacy protection, intellectual property frameworks, and regulatory structures that govern what people do with cryptocurrency, even if only to say "nobody can torture you for your private keys."

Somebody needs to make bitcoin easy, accessible, and safe for your average person to use.

In doing so, that "somebody" has many, many opportunities to screw you over.

Isn't that what Wall Street does best?

Central Bank Digital Currencies are an Extension of the Flaws of Fiat Currencies

4

Written by Sylvain Saurel

The Central Bank Digital Currencies (CBDC) have been getting more and more attention in recent weeks. The Chinese digital yuan project, the famous e-RMB, is moving at high speed. The e-RMB project is so advanced that it is already being tested by the general public in four major Chinese cities.

China wants to take advantage of the first-mover advantage in its new battlefield to maximize its chances to break the U.S. dollar's monopoly on the world monetary and financial system.

Under these conditions, it is easy to understand why Xi Jinping has made e-RMB a national priority since October 2019. Everything indicates that China will be the first country in the world to officially launch its CBDC.

The Race to Create Their Own CBDC has Begun for Central Banks

Behind China, many countries are announcing that they are working on their project. Many feel that something is happening at the digital level, and China should not be left alone on this new battlefield that could become strategic in the future.

I recently saw a survey indicating that 80% of the world's major central banks were thinking about implementing their CBDC.

The European Union is advancing on a digital euro project with tests planned for 2021. Japan and South Korea are already well advanced on their project. Finally, the United States is still hesitating. However, discussions within the Fed are becoming more and more intense.

Some American Senators are pushing in this direction, explaining that the delay currently taken by the United States could have unfortunate consequences in the years to come.

Sooner or later, the United States will create a digital version of the U.S. dollar. This is in line with history, and they will have no choice but to follow the movement initiated by Xi Jinping's China.

The Challenge Will Then be to Convince Citizens to Use These CBDC

When each country, or economic zone in the case of the European Union, will have its own digital currency, the challenge will be to convince citizens to use these CBDC. Of course, a major education program will have to be conducted in each of these countries.

The majority of citizens are far from being ready to use digital currencies at present. The educational work that will be done will benefit Bitcoin sooner or later. You can see this as a Trojan horse. Indeed, once educated in the use of these digital currencies, citizens will probably go and test alternatives that are more protective of their privacy.

Bitcoin will inevitably be the number one solution when they decide to try another more respectful way.

You may wonder how I can be so sure of that. It's quite legitimate. So I'll explain why, in a future world where everything will be digital, Bitcoin will always remain a separate species.

These CBDC Will Continue to Rely on Centralized Systems

First of all, these CBDC will all be centralized. They will only be a digital version of the existing fiat currencies. In fact, current fiat currencies are already digital when you think about it since more than 85% of the world's financial transactions are done digitally today.

Therefore, users will face the same problems that the centralization of the current system creates.

Cash allows people to have a minimum of privacy when it comes to money. Once these CBDC are widespread, the goal is to eliminate cash. Under these conditions, all financial transactions carried out will be even better monitored.

This will endanger the privacy rights of hundreds of millions of people around the world more than ever before.

With the current system, citizens could at any time see what they have in their bank accounts being confiscated.

Fortunately, cash allowed them to have some security of a property. With these CBDC, the possibilities for confiscation will be total. This will create even greater problems of censorship. As all transactions can be monitored in real-time even more simply, governments may be tempted to sanction citizens for purely arbitrary reasons.

It is no coincidence that China also sees its e-RMB project as the last missing piece in the surveillance society it has been building for years with the Chinese social credit system.

These CBDC Will Not Allow Citizens to Have an Efficient Borderless Payment System

Bitcoin is a borderless electronic payment system. You can easily and quickly send BTC to the other side of the world to a friend in complete security. It takes about ten minutes and the fees are minimal. Besides, Bitcoin works continuously which is an incredible added value compared to the current banking system.

With these CBDC, the problems of the current banking system will only be transposed into a 100% digital world. No benefit for the users.

The permissionless side of Bitcoin is an incredible advantage for its users. Bitcoin belongs to everyone and has no leader. Everyone can use their money as they wish on their network. With these CBDC, this possibility will not exist.

These CBDC will belong to the governments that will authorize you to use them. Here again, we find the same flaws as with the current monetary and financial system.

Endless Inflation of Fiat Currencies Will Continue With These CBDC

Triggered by the coronavirus pandemic, the economic crisis of 2020 has once again brought to light the only strategy that central banks knew in times of crisis: to print more and more fiat money out of thin air while lowering interest rates to zero or even in the negative territory such as the European Central Bank for example.

More than 12,000 billion dollars have already been printed since March 2020 by the world's major central banks.

This propensity to print more and more fiat money is the original sin of today's monetary and financial system.

From this original sin flow seven deadly sins that will cause the loss of this system sooner or later.

Such inflation of the world money supply has dramatic consequences for the people. It constantly devalues what the poorest possess. With no other option than cash in these times of great monetary inflation, the hundreds of millions of the poorest people continue to bear the brunt of the injustice of this system.

With CBDC, this will not change. I would even say that it will facilitate the work of the Fed and all other central banks. Increasing the M2 Money Stock will be even simpler. When we see what the Fed has already done since March 2020 with the current system, we can fear the return.

This great monetary inflation that we are experiencing highlights more than ever the unique characteristics of Bitcoin. The supply of Bitcoin is hard-capped at 21 million units and its programmatic monetary policy is gradually reducing the issuance of new Bitcoins.

By 2140, when all Bitcoins have been issued, the world will have to make do with the supply already in circulation. It is no coincidence that Bitcoin is said to be the scarcest human invention in the world.

Bitcoin is even the hardest money in the world.

Conclusion

Bitcoin's unique monetary attributes are supported by its disruptive Blockchain technology. This supports the Bitcoin revolution that aims to build a fairer world for all in terms of money in the future.

While some may see CBDC as competitors for Bitcoin in the years to come, I hope I have shown you that Bitcoin's decentralization and unique monetary policy will always make it a species apart. Bitcoin plays in a category of its own.

Bitcoin is the money of the people supported by the people, which the CBDC can never be. They are merely a digital version of the fiat currencies at the heart of a monetary and financial system that is flawed and not fixable.

These CBDC represent nothing innovative, and sooner or later the people will know how to make the right choice by opting for Bitcoin.

Why Bitcoin Matters to Millennials

5

Written by Mark Helfman

I had conceived of this article as a Ted talk. Maybe one day, if I'm lucky.

For now, I'm publishing it here. It's something that I've been thinking about for a while and I feel compelled to put it out there. Somebody needs to say it—why not me?

This article is about why bitcoin means so much to the Millennial generation.

In one sense, it's a defense of bitcoin from people who can't understand why it has any value or why we would spend our time and money on it.

In another sense, it's a reflection on Millennials—our history and values.

As the first generation that came of age in the 21st Century, we have a unique perspective that's worth exploring.

It's Not About the Money

When I talk about cryptocurrency, I get the same questions again and again:

"What's bitcoin backed by? Should I buy it? Do you think its price will go up?"

Instead of opening up conversations about the nature of wealth, finance, privacy, and government, it always comes back to money.

Yes, bitcoin is money, but it's also *more* than money.

Bitcoin allows free, public access to financial networks on-demand, without restriction, anywhere, any time.

As you read this article, developers and businesses have already started using this open, permissionless payment protocol to build real products and services that lower the costs of doing business, get rid of barriers that keep entrepreneurs from getting loans and taking payments, and make money easy, cheap, and convenient for people who don't have banks or trustworthy governments.

This is a luxury we take for granted, but for many people, it's a privilege normally reserved for a select few.

And this matters because . . .

. . . because it's all happening organically, without any government involvement or subsidy. It's happening *despite* the opposition of powerful elites.

To understand why that's so important, you need to understand the story of the Millennial generation. You need to understand why so many of us have lost trust and faith in our leaders.

The Millennial generation entered adulthood during a recession followed by a war, followed by another war, followed by a financial collapse, followed by a massive bailout of wealthy people and large banks, followed by a massive tax cut for the rich and powerful, followed by more debt, pandemic disease, and another recession.

We have grown up in the midst of the greatest transfer of wealth from the middle class to the rich, while the poor remain poor—and all of it financed with debt and subsidized by workers.

Everybody told us that the stock market always goes up, a house is the key to wealth, and education is the key to a great job.

Then the stock market crashed, houses got too expensive to buy, entry-level requirements shot up, and entry-level pay went down.

When we skimp on savings to pay our landlords and pay back our student loans, it's not because we're irresponsible. We're trying to dig out of the hole we've been put in.

Other generations mock our lifestyle choices. They build their political power and financial fortunes from our time, talent, and ingenuity, then blame us for their problems.

How many things have millennials killed so far?

Google it.

That's the story of the Millennial generation.

The System Has Failed

In the U.S., our generation has seen our government pass a budget on time ONCE. Twice, it didn't pass a budget at all. And never have any of those budgets been balanced.

We have gone from a war against terror to war against trade to a war against ballots. From a president who cheated on his wife to a president who cheated on his country.

We have seated three presidents, but only one got into office by winning the majority of the votes.

Our electoral experience starts with dangling chads and ends with Russian interference. Our votes are gerrymandered, cherry-picked, and where there are too many people affiliated with one party or another, ignored.

Our politicians rig the system in favor of their own interests or against their opponents' interests while the voters lavish votes and money on whichever party seems to say the right things at the right time. Each party conspires to undermine the Constitution in one way or another, the only question being how, how much, and how openly.

We're being pummeled by political parties while we argue about which fist we prefer, the left or the right.

Some say we should be more politically active, but the moment we run for office, they call us young, inexperienced, naive, stupid, reckless, and ignorant.

That's the story of the Millennial generation.

Leaders, Where Are Ye?

Instead of bringing us together, our leaders insist on pitting us against each other.

Main Street against Wall Street. Wall Street against farmers. Farmers against coal miners. Coal miners against high school students. High school students against gun makers. Gun makers against black people. Black people against the police.

Our police do everything right almost every moment of every day. They save us from fear, terror, and death, preserve safety and order for millions, and improve our communities in many ways—but one horrible mistake, one incident out of millions, and they're despised.

A knee taken on a sideline destroys a man's livelihood. A knee taken on a neck destroys another man's life.

And still, everybody misses the point.

When we advocate for what matters, we aren't advocating for the oppressed. We're advocating for our friends and colleagues. Our neighbors and family. Our coaches and teachers.

When we say we want a country where we can run for public office without fear our president will bribe a foreign country to frame our children for fake crimes simply because we campaign against him, it's not because we're mad about the last election. It's because we care about the integrity of our government and the rule of law.

One day, principles will matter more than votes, money, and "earned media." Until then, it's feed the beast, wag the dog, bread and circus.

That's the story of the Millennial generation.

Fakes, Frauds, and Failures

Our leaders have the power to stop all of this, but they don't. Instead, they use their power to incite more hatred and violence.

They claim to serve those cherished institutions that have bound us together for generations, but they use those same institutions to rip us apart.

Social media platforms claim to offer a voice to the voiceless, but their algorithms spread only the most salacious and aggressive messages while hiding anything they consider mundane.

Outrage is bought and sold. Controversy is mass-produced—better yet when it's perfectly tailored for a fundraising letter or a campaign speech.

We live in a world of fake news spread by fake profiles where the only thing that's real Is the artificial intelligence that created them. Fact is fiction and Facebook is reality.

Anybody can ignite a culture war with 280 characters typed into a cellphone and posted on an electronic bulletin board, all the while, solving nothing. All that anger and nothing changes.

That's the story of the Millennial generation.

Bitcoin Fixes This

So what's that got to do with Bitcoin? Why does bitcoin matter to us?

Because bitcoin doesn't care about any of that.

It doesn't care where you live, how much you make, or who you marry.

Interest rates too high or too low? Doesn't affect bitcoin.

Data breaches? Bitcoin doesn't use your personal data.

Banking collapse? Bitcoin doesn't need banks.

Corporate bailouts? Welfare state? Immigration? Tariffs? Trade wars? COVID-19?

None of that matters to bitcoin. It's just an algorithm. A computer program. It doesn't get strong or weak based on what political party is in power or who sits on the Federal Reserve Board.

You don't lose your bitcoin when you get fired or when your company goes out of business like you lose your healthcare or your pension.

You can't bribe bitcoin.

Bitcoin doesn't care how old you are, how young you are, how rich you are, how poor you are, whether you have a job, or whether you don't have a job.

It doesn't treat you differently if you're wearing a suit instead of a t-shirt, or because you share the same skin color as the person you're doing business with.

It doesn't lavish money on big businesses while throwing crumbs of fortune at the small ones.

It will never sell your personal information to a company or government that will use it against you.

Bitcoin will never do any of those things—it's programmed not to.

Finally, a Choice

For all the things I've said, I don't think by and large this generation dislikes banks and governments and powerful institutions. Most of us aren't rallying or lobbying about that.

It's not something we really think about much. We're doing what every other generation does—buying, saving, borrowing, working, and living.

We use all the conventional "things" you use in a developed economy. 90% of us have a checking account. If we need a loan, we get a loan. If we need a credit card, we get a credit card. If we want to buy something, we buy something.

There's no "anti" here. We use whatever's convenient. Whatever works often enough that we're OK with.

The difference is, we have bitcoin. With bitcoin, we have a choice. The choice is:

- Trust a steady, mathematically proven, immutable process that no person can control or manipulate, that always gives you the result you expect.

- Trust a government and financial system that have always failed us.

Get Your Head Out of the Bubbles

I realize previous generations don't see this as a choice. They just think it's silly.

They would never put their fate in the hands of a formula, a network of strangers, an asset that only exists in pixels and lines of computer code.

I get it, and it's a valid point.

My problem is, the conversation keeps us from getting into that.

The conversation begins and ends with "that's YOUR thing" and in between, you get a whole bunch of people who can't get over the idea that bitcoins are not backed by anything.

"It's ridiculous! It's a bubble! I don't get it. It doesn't solve any problems. You can't make money from it. Nobody uses it. It always crashes. It's slow and expensive. Rat poison squared. Stupid internet money. I'm buying gold."

Every blockchain project that fails is met with "I told you so." Every time the price of bitcoin goes down, "I told you so."

That's the story of the Millennial generation. "I told you so."

It started with the dot-com bust, a tragedy we had nothing to do with.

In 2000, after the bubble burst, all we heard was "the internet was a stupid, terrible idea, a big bubble, it doesn't solve any problems, you can't make any money from it, nobody uses it. Stupid. Told you so."

Meanwhile, our hopes of a great paying job disappeared when the market collapsed.

A few years later, we climbed out of that hole and started putting money into the one asset that everybody said was the BEST investment you could make, the key to financial freedom, the bedrock of the American Dream — a house.

And what happened?

The housing market collapsed. In some places, it hasn't recovered.

Which One Was The Bubble?

Today, we may not own houses as much as older generations, but we use the internet. It's everywhere. It's everything. It's what kept the world moving when the pandemic hit. It saved millions of people's lives.

It's the first big innovation we ever had and it's the innovation that molded us.

And everybody said it was a stupid, terrible idea. Go buy a house.

So when they say bitcoin is a stupid, terrible idea, yet you're reading this article because of the internet, from a guy who wrote two books that would have been impossible to research without the internet, distributed through a network that would not have existed without the internet, publishing an article that you can view for free everywhere in the world because of the internet, to an audience that would have never known about it without the internet, with as much authenticity and credibility as the most famous expert in this field . . . do you think we're going to believe those people who say "I told you so" about bitcoin?

No. That's not the story of the Millennial generation. We have a choice. Our choice is:

Trust a steady, mathematically proven, immutable process that no person can control or manipulate, that always gives you the result you expect.

Trust a government and financial system that have always failed us.

Which would you choose?

Bitcoin & Millennial's - Why it May or May Not Matter

6

Written by Thomas Wolf

I found Mark Helfman's recent article *"Why Bitcoin Matters to Millennials"* to be thought-provoking and, for the most part, highly relatable. Below is my response to his article intended to help communicate ideas between team community writers and readers.

"Share your knowledge. It's a way to achieve immortality." -*Dalai Lama*

I was born in September of 1989, which makes me a millennial. It seems I am the perfect candidate to respond to Helfman's article as it is focused on how Bitcoin means a great deal to the Millennial generation, which currently represents people born between the years 1981 and 1996. However, the term is loosely used for younger generations - I fall right in the middle of this timeline demographic. I'm not proud of it, but I try to make a difference for those above and below my age groups.

I agree with Mark on several counts, starting with Bitcoin, which is to say the software engineering behind Bitcoin and blockchain technology is more than money, despite being a monetary currency of value. I've also noticed the laissez-faire attitude of non-government involvement in cryptocurrency markets, a term generally noted in economics to represent systems that function better with minimal to zero government involvement. I can most definitely relate to growing up in a time where "*the greatest transfer of wealth from the middle class to the rich, while the poor remain poor,*" as Helfman stated.

There is then the statement that "*the system has failed,*" to which I reply, the system never succeeded for the people in the first place. Politics have begun to become integrally tied to cryptocurrencies as they are now a significant market, the only market that has more than recovered from the COVID-19 pandemic crashes. War, religion, race, sexual identity - these have all suddenly started to affect the marketplace through civil and corporate unrest.

The government did fail us, but the government is changing. Bitcoin regulation has appeared. Services such as Paypal and Robinhood want to hold your assets in a form where you do not have access to your private and public keys and do not own them. Many countries have passed laws in relation to cryptocurrencies already, so government involvement is inevitable. I've seen other writers argue that even with the rise in DeFi currencies and exchanges, nothing is genuinely decentralized as that money can be followed until it reaches a centralized currency for an exchange to FIAT.

I argue that it was the previous generations that allowed this to happen by presenting us with a rigged system to vote for the lesser of two evils; taking the blame for trying to clean up my elders' mistakes is not on my to-do list. I cannot take responsibility for how the masses raised their kids when I was busy raising myself. This new frontier is one small beam of hope for our crumbled economy here in the USA. China has already demonstrated that renewables are more profitable than traditional energy sources after sticking to the Paris Climate Accord signed with former President Obama.

I feel that leaders are in a controversial spot right now because there is such strong opposition in massive numbers on both sides; it's challenging to maintain any form of conservative stance without looking undeserving to one party or the other. It is times like these, I am proud to be independent and prefer to evaluate individual instances in my own way, on my own time. To create a better sense of diversity, I believe we need to seek the truth; the truth is the only thing that can genuinely silence ignorance. Honesty and knowledge give people the power to make informed decisions that will likely benefit one another instead of having them at each other's necks on every issue, particularly those related to money.

"Bitcoin fixes this" is a loaded statement after discussing issues like failed economies, fraudulent and failed leadership in government, and the general mistrust of authoritative political figures in young people. I argue that blockchain technology offers many applications for solving political system issues, such as blockchain in US voting polls, to create an ultimately unhackable and accountable system. For me, one recent area of interest is blockchain tech within agroecological systems to help fairly manage resources on a smaller local scale - a subject I plan to write on further.

However, there are areas in which I beg to differ; while Bitcoin has many attractive features, it is weak compared to modern derivations and evolutionary results such as EOS or Nano. The statements you've made about Bitcoin being programmed not to do many things that most FIAT corporations enforce, as well as the context around it being software that does not care, can be equally applied to other cryptocurrencies, if not more so around ones that have been developed with these interests in mind, and improvements over the slow and power-extensive Bitcoin. I waited all night for a few BTC confirmations to clear yesterday; they should only take 10 minutes, and that is a ridiculous amount of time to wait for a crypto asset with a transaction fee; this is 2020, after all.

I feel your point falls flat on its face when you have to consider cashing out of your cryptocurrencies and staying within the law. The moment it hits your bank account from a centralized exchange, you are back to playing the man's game and subject to the same laws as everyone else. While this is currently a legal gray area, it will not remain that way. The software won't change - but the laws on using the software will, and then the software will have to be developed (altered) to adapt to those laws and the holders' new priority demands. Alternatively, software would be developed to break them.

My conclusion is that we agree on a great many things, my friend. But I also think that your views are short-sighted when it comes to the level of control that will be in place over "gray markets." To directly answer your question, I will always trust a system that is mathematically proven and entirely in my power with no third party required. Still, one has to be always conscious of the fact that at any time, laws and market changes can render your assets to be worthless. You have to be ready to evolve with cryptocurrencies as a whole, as I've learned all profits in this racket come from having your money in the right places at the right times.

Questions and comments are always welcome. I look forward to your thoughts, Mark!

Axie Infinity Empowers Players with Axie Infinity Shards (AXS) Tokens

7

Written by William Peaster

Traditionally, games have been siloed and owned by companies. But what if games were open, community-owned, and had native economies that could enrich plays and developers alike? What if you could farm game assets and sell or trade them on the open market?

That's where an NFT-based Ethereum game like Axie Infinity comes in. Influenced by Pokémon, Axie Infinity is a game where you can collect, battle, and breed digital pets called Axies. The game notably takes a "play-to-earn" format, in which players battle in player vs. player (PvP) and player vs. environment (PvE) matchups to earn gaming assets like Small Love Potions (SLP).

These tokens can be sold through trading protocol Uniswap for ETH (note: we've even started to see some players shard rare Axies into fractionalized pieces so they can be traded freely on Uniswap and beyond). Alternatively, SLP tokens can be used to breed new Axies, which leads to players farming SLP, breeding and selling new Axies, and growing the game's economy all the while.

So that's the basic gist of the Axie Infinity project. Yet the game's builders just kicked things into high gear in unveiling the new AXS governance token, which will be used to shift Axie Infinity into a truly community-owned project that rewards its stakeholders.

Meet AXS

On Monday, October 26th, the Axie Infinity team announced it would be launching its new AXS (Axie Infinity Shards) token through Binance Launchpad, crypto exchange giant Binance's initial exchange offering (IEO) arm.

An ERC-20 token, AXS will function as the governance token of the game: holders will be able to stake their AXS tokens in order to participate in votes to steer the Axie Infinity project, e.g. the game's coming to Community Treasury. Moreover, these holders will earn weekly rewards for staking their AXS and participating in governance. AXS will be purchasable on exchanges like Uniswap or earnable by playing Axie Infinity itself.

And just like how Uniswap retroactively rewarded past Uniswap users with UNI when the decentralized exchange launched its governance token, Axie Infinity is also going to be awarding its players with AXS depending on the extent of their past interactions with the game.

As for the Binance Launchpad sale, AXS will have an initial price of $0.1 USD per token and an initial circulating supply of 59,985,000 tokens (the token's total supply will eventually be 270 million).

Why it Matters

Axie Infinity isn't just one of Ethereum's most popular games right: it's also one of the network's most popular apps, period. According to the curated dapp directory State of the DApps, Axie Infinity is currently the fourth-largest Ethereum app in notching nearly 3,000 users over the past 24 hours.

That said, for Axie Infinity to open up its game to its community of players with AXS represents a bold, high-profile early experiment in play-to-earn gaming that could influence other games to follow similar paths in the years ahead. There may come a time, then, where we look back and think it was strange there was ever a time when play-to-earn gaming wasn't a thing.

If that does ever end up being the case, then Axie Infinity will have been a key early trailblazer that paved the way for the future of gaming. In the meantime, the community bides its time until the Axie Infinity Community Treasury goes live next year. Eventually, AXS holders will be able to vote on how this fund gets used to foster the Axie Infinity ecosystem, and the fund itself will be grown by the game's breeding and marketplace fees.

Enterprise Blockchain Adoption, 2021?

8

Written by Scott Owen

Anyone who has worked in Enterprise would undoubtedly have heard of or used the technology analyst firm Gartner. Firms like Gartner are famous for their research, models, and IT conferences aimed at company C-Suite or Executive Leadership teams. Gartner surveys and advises 1000s of CIOs globally each year, providing research and strategy advice relating to technology, people, process, vendors, and more. To many, Gartner analysts are comparable to Oracles or prediction machines, with in-depth knowledge and great wisdom. Their predictions typically come to fruition, not always precisely, but within a degree or two of accuracy.

Whatever your stance on Gartner, or similar analysts there is one certainty. Technology leaders listen to Gartner and respect their advice! The firm functions as a sounding board, saving leaders time whilst acting as an insurance policy in the future. If Gartner ticks your initiative, there's a reasonable probability you'll get approval, funding, and if things go south, you're less likely to wind up unemployed. With technology leaders relying on Gartner to reduce their risk, rest assured, Gartner will take a conservative route and recommend things with a high likelihood of success (or a low chance of failure). They are unlikely to tarnish their reputation by sticking their neck out for fresh out of the gate unproven technology. They will wait for it to be proven or deemed *safe* for Enterprise.

People being people, however, are frequently distracted by the shiny and new. A novel solution has an attractiveness; an allure which makes it irresistible to technology nerds alike. For this, Gartner invented the *Hype Cycle*. The model is their way of introducing new technologies to executives and senior leaders before having to commit.

"When new technologies make bold promises, how do you discern the hype from what's commercially viable? And when will such claims pay off, if at all? Gartner Hype Cycles provide a graphic representation of the maturity and adoption of technologies and applications, and how they are potentially relevant to solving real business problems and exploiting new opportunities."

In 2016 Blockchain technology made its debut with a 5 to 10-year time horizon. In recent years it has slid into the Trough of Disillusionment where it now sits. In recent times, Gartner broke Blockchain out into its individualized Hype Cycle, demonstrating just how much is currently under development.

The question is in a post COVID world, where does Gartner stand? Analyst after analyst and everyone I speak with agrees the global pandemic accelerated secular trends which before were 5-10 years out. From one day to the next, those computer bound people like myself, working from home became the norm. Video and remote collaboration became pervasive overnight.

A shift towards digitization is no longer an option; it's mandatory. Companies globally are working day in and out to restructure their businesses, their workforce, and balance sheets to keep pace in the digital world. Those who fail to shift their revenues online risk insolvency or will flounder as the landscape gets extra choppy over the years ahead.

Analysts like Gartner had predicted Crypto and Blockchain would not reach the Slope of Enlightenment until 2025. However, have recent moves shifted the thought processes?

The chances of a company like MicroStrategy, which is Nasdaq listed, purchasing USD 425M of Bitcoin to protect its company treasury to avoid currency debasement were near zero before 2020. Or the probability of Square, the global payments provider buying USD 50M of BTC to stick on their balance sheet was low at best. Or the likelihood of Mode committing 10% of their cash reserves to Bitcoin would be improbable. The question for me is who's next?

Will Google, Facebook, or another tech giant adopt the Bitcoin Standard?

In fairness to Gartner, no one predicted 2020. Now it is happening (or happened), companies are reacting and moving forward on their timelines. Recently PayPal jumped into Crypto and

Blockchain by opening its merchant network to buy and sell Crypto within its system. The move brings even greater visibility and attention, making it more and more difficult for analyst firms to ignore.

This shift was further cemented by the IMF convening an online conference on Central Bank Digital Currencies (CBDCs) and a new Bretton Woods moment. Furthermore, the World Economic Forum is calling for a Great Reset, in which technology like Blockchain is at the center of its master plan.

How about Google, one of the world's largest tech companies becoming a Block Producer candidate on the EOS Public Blockchain.

Blockchain and Cryptocurrency advocates understood the value of the technology years ago, and now the world is embracing and catching up. Institutions and significant investors are coming. The case for Bitcoin as a Life Raft from the current monetary system was made by Raoul Pal, the founder & CEO of Real Vision.

Tech Analysts Can No Longer Ignore Crypto and Blockchain.

Now for a prediction of my own. I think Gartner and analyst firms alike, will begin recommending CIOs and Executive Leaders of Enterprise companies move Blockchain projects out of Proof of Concept to full development in 2021. The pandemic has accelerated many slow-moving trends, in some cases transporting us forward years in only a few short months.

A recent Gartner survey suggested two-thirds of ANZ CIOs are uninterested in Blockchain. On the flip side, around 40 percent have an interest, and if firms like Gartner recommend in earnest, it's likely to drive enterprise confidence and adoption higher. I can't wait to see how things progress.

Can Satoshi Nakamoto, Bitcoin's Creator, Become the Richest Person in the World?

9

Written by Sylvain Saurel

Little by little, the Bitcoin price is getting closer to more heights not seen since the end of 2017. Last week, the Bitcoin price reached $13K for the first time since June 2019.

Bitcoin now has a market cap of around $240 billion.

Carried away by their enthusiasm, some have therefore begun to imagine the start of a Bull Run that would bring the Bitcoin price to $20K by the end of 2020.

I don't know for sure if the price of Bitcoin will have reached $20K by the end of 2020. However, if it does not, it will only be partly postponed, as the long-awaited Bull Run will take place during the year 2021. It's no longer a question of if, but rather a question of when from my point of view, and obviously that of all Bitcoiners.

Bitcoin's Creator Could Become Richer than Jeff Bezos in the Future

From then on, many start to dream and wonder whether Satoshi Nakamoto, the mysterious creator of Bitcoin, could eventually become the richest person on the planet.

I guess people who ask this question reason in terms of estimated personal wealth. For example, Jeff Bezos is the richest man on the planet with an estimated fortune of $188 billion. His fortune is made up of 90% Amazon shares.

This can vary greatly from one day to the next depending of course on changes in Amazon's share price. Thus, in the heart of the summer of 2020, Jeff Bezos' fortune increased by more than 10 billion dollars in a single day for example.

Jeff Bezos even became the richest man in history in the summer of 2020 with an estimated fortune of over $200 billion.

Amazon has a market cap of 1,610 billion dollars while we are on October 23, 2020. The upward trend of this market cap suggests that Jeff Bezos has the potential to reach $250 billion in personal wealth in the future.

If We Know the Exact Fortune of Jeff Bezos, that of Satoshi Nakamoto Must Be Estimated

Now, let's take a closer look at the mysterious Satoshi Nakamoto. First of all, keep in mind that we don't know whether Satoshi Nakamoto is a physical person or an entity composed of several people.

Nevertheless, let us imagine that he is a physical person.

Satoshi Nakamoto would possess between 800,000 and 1 million BTC. This naturally makes him one of the largest BTC owners in the world. I invite you to look at the current distribution of the Bitcoin circulating supply to better measure this:

As of October 26, 2020, the price of Bitcoin is $13K.

Satoshi Nakamoto's current wealth in US dollars would therefore be between $10,400,000,000 and $13,000,000,000.

We are still far from Jeff Bezos' fortune which is currently $188,600,000,000.

Satoshi Nakamoto Will Benefit from the Greater Upside Potential of the Bitcoin Price

The advantage that Satoshi Nakamoto could have is the incredible growth potential of the Bitcoin price for the future. I personally see two main reasons why the price of Bitcoin will reach the million-dollar mark within 20 years:

1. The law of supply and demand. The demand for Bitcoin will continue to increase while the supply of new Bitcoins will become increasingly scarce. Following the principles of the law of supply and demand, the Bitcoin price is expected to rise sharply.

2. The endless inflation of the U.S. dollar. The year 2020 will have shown that the Fed considered more than ever that printing more and more U.S. dollars out of thin air was the solution. In reality, this is the root of the problem, but the Fed continues to resort to this remedy, which only makes the situation worse in the long run. In any case, this endless inflation of the U.S. dollar and other fiat currencies will benefit the Bitcoin price in the years to come.

If Bitcoin were to reach a price of $1 million in the future, then Satoshi Nakamoto's fortune would be between $800,000,000,000 and $1,000,000,000,000.

Satoshi Nakamoto's fortune would then be four to five times greater than that of Jeff Bezos' current fortune. Of course, by then, Jeff Bezos' fortune will have had time to increase since it seems obvious that Amazon's market cap will increase again.

But it is not certain that Amazon's upside potential is sufficient to compete with Bitcoin's in the future.

Conclusion

There is a real chance that Satoshi Nakamoto, as tempted as he or she is a physical person, may one day surpass Jeff Bezos' fortune by becoming the first trillionaire in history.

In the meantime, a very successful entrepreneur like Elon Musk may have passed Jeff Bezos, because he has the advantage of having founded several high-potential companies like SpaceX and Tesla.

Gamifying Crypto Ecosystem: Serious Games, Gamification & Game-Based Learning

10

Written by Alex Cuevas

Providing tools that facilitate both understanding and interaction of the broader public with the crypto ecosystem is one of the key aspects in order to achieve mass adoption.

In this article, we will discuss in a deeper way some learning concepts, currently subject to confusion, such as serious games and gamification, as well as reviewing certain available platforms within the blockchain industry.

I think it would be advisable to approach strategies according to different standards, such as previous experience, occupational background, or personal skills and interests, in order to reach most people in our society, focusing on the practical aspects which will allow them to determine if they are onboard into a new economy, or on the contrary, feel more confident in their current status.

Background

Serious games, despite its popularity, is a term conceived over 40 years ago by the German researcher Clark C. Abt, when he discussed the possibility of incorporating games whose main goal is teaching instead of entertaining.

On the other hand, gamification is considerably more ancient, as the aim is to move gaming elements into unfunny scenarios.

Probably one of the most curious examples of gamification throughout history was created by Angelique du Coudray, a midwife, who in the 18th century designed "*The Machine*", a lifesize obstetrical mannequin for practicing mock births.

She trained hundreds of women in France on the practical implications, saving literally thousands of lives. Subsequently, based on her success, hundreds of similar models were created and distributed, enabling the same teaching model to be expanded even to more countries.

Gamification

In spite of a few positive cases, I think most organizations trying to include gamification in their structures are just scraping on the surface, without moving further than the usual PBL systems (points, badges, and leaderboards).

These strategies might be effective in the short term, but in the medium and long term users will tend to feel very bored, simply for the fact they want to enjoy playing.

Nevertheless, within the crypto ecosystem, some projects have successfully taken gamification to a new level with exciting projects, such as Kallium, an official Banano (BAN) crypto wallet.

Each aspect of the design plays a part in allowing a better understanding of crypto abstract concepts. In the educational sphere, particularly for children, environments like Kallium can help to massify the comprehension of cryptocurrencies.

Moreover, blockchain gamification can exist in real-life scenarios, for example, some companies design crypto coin wallets with the shape of physical coins, which may seem crazy but may help first-time users to introduce them to the crypto world.

Serious Games

Serious games, on the contrary to bringing game aspects into the real world, are trying to take real-life simulators to the gaming environment. In this case, there are genuine masterpieces that we cannot overlook in the crypto ecosystem.

Each aspect of the design plays a part in allowing a better understanding of crypto abstract concepts. In the educational sphere, particularly for children, environments like Kallium can help to massify the comprehension of cryptocurrencies.

Moreover, blockchain gamification can exist in real-life scenarios, for example, some companies design crypto coin wallets with the shape of physical coins, which may seem crazy but may help first-time users to introduce them to the crypto world.

Blockchain Games

Finally, many games are built using blockchain technology, so they are considered just games. In this case, it is not necessary to have any relation to blockchain game-based learning, simply the data treatment is conducted using blockchains.

To conclude, I would like to remind you of a sentence that Plato said more than 2,000 years ago. "*You can discover more about a person in an hour of play than a year of conversation*", so will we be able to find tools to make crypto education fun? Feel free to express your opinion by commenting, thanks for your support!

What On Earth Is Going On In DeFi?

11

Written by Thomas Daniel

All fall, we've heard the hype. Everyone is excited about DeFi! Well, at least until they aren't anymore. I don't necessarily see what the buzz is about. Sure, you're working with stablecoins for the most part and they don't fluctuate in terms of value the way other crypto tokens can, but it's a bull market even despite all of the cataclysmic ongoings in the world today. It isn't the most rewarding bull market thus far, and who knows where we might be if the pandemic hadn't struck, but despite all of these external bearish indicators, we're still looking at growth. That signifies strength.

But there are some strange things going on. For one, the transactions cost too much to make small plays in DeFi. The only person who should be putting money into DeFi is someone who has a substantial amount of it and is willing to either get creative and take risks to increase the returns or someone who is comfortable at a low return on the investment.

This hasn't stopped the hype from exploding, and I must confess that I do not pretend to understand what's going on here. I've done a few experiments, though, and in this article, I'll explain what happened and try to wrap my head around this market. Any and all comments are welcome, as it's confusing stuff and I'm definitely still just starting out too.

DeFi Hype Hits Coinbase

Earlier, Coinbase paid me about $20 in 3 different cryptocurrencies to lock $3 of USDC into a Compound smart contract that will earn me 2.2% APY on that money. Shrugging, I followed the steps to earn the $10 COMP prize that I was promised, and I was pleasantly surprised to see gas prices dip dramatically between the time Coinbase estimated my gas fee and sent me ETH to make the transaction with and the transaction's actual execution moment, which resulted in about $6 in ETH remaining in my wallet, an unintended bonus perhaps.

The transaction fee for my $3 smart contract deposit was only about $1, ultimately, which would seem reasonable aside from the fact that the money will have to remain in the contract for the rest of my working life to earn back the transaction fee.

My experience with Curve was very similar. I believe I spent about $80 in transaction fees to make $2 on a small investment of $150. I just wanted to try it out, compare it against my inflated bag of ATOM tokens, and see what all the buzz was about. My conclusion? Not much.

The UIs on most of these things are clunky and old school. Curve in particular reminds me of Windows 95 or something, but the main goal is to make money, so who cares what it looks like? Assuming the transaction fees would have been the same if my deposit had been $15,000 instead of simply $150, I'd have made $200 by exercising the transaction along the same lines as I did. From this $200, we subtract the $80 in fees, for a tidy profit of $120 for loaning out my $15,000 for about six weeks. This rate of return is better than what we can hope for in a savings account but not as good as the stock market on average.

Scenarios

Why is DeFi so hyped, if it's not relevant to every situation? I have a few theories here. And, believe it or not, it does actually seem to be a real financial vehicle worth investigating. I'm still not sure that ordinary people will ever hold enough BTC to buy a car and wait for the appreciation to pay the note off, but for early adopters, there are definitely some financial hacks that can help us make money in the markets and increase our spending power courtesy of DeFi protocols.

The first theory is that DeFi is a place to keep earning between big, well-thought-out moves in other crypto tokens. If you hold a bunch of ATOMs and actually sell them when the price has tripled, you may not see another growth opportunity in the market right away even if it's a bull market. So you throw the profits into DeFi smart contracts and wait for the price to come back to earth, completely safe from any asset depreciation no matter how rocky the market gets. Then, when it's at the bottom, you close out of the liquidity pools and putting the money back into the market to ride the escalator up again.

I wish I were a solid enough trader to continuously do this sort of thing. I hit about 50/50 on my day trades and switched to months-trading, and then became sort of a believer in the Cosmos ecosystem and am simply waiting for that to pan out in the mid to long term.

The second theory is that crypto-based lending is allowing people to overleverage their assets by taking out loans against them. I've used Nexo Wallet to do this a few times since August and I'm 0/3 or 4 at the moment, but my most recent play may turn out fairly well. I keep thinking I see short-term plays and staking my ETH and BTC to borrow 50% of its value and then losing some money on whatever I buy with it. Sometimes, this is my goal. I turn the USDC I get from Nexo into XLM, then head over to Stellarterm to buy TERN with it, then spend the TERN from my Blockcard to buy things I need.

Ideally, the price of BTC and ETH would rise while the loan was out, and I would be able to repay the loan by selling only a portion of the crypto that I staked against it when I took it out. One good beat there, say if BTC hits $20k again, could really help me out. I could even begin purchasing BTC and ETH with the leveraged loans and simply ratchet up my exposure to the assets. If you have 1BTC, without selling your BTC, you can actually acquire a second BTC via Nexo's lending protocol.

How this works:
You can borrow about half your money each time, and if you do it enough times, you end up with about 1 BTC you started with and a second that you borrowed money at a set price to purchase. It probably takes half a dozen transactions, but you can actually gain

double exposure to an asset. The problem is that you risk liquidation if the price dips too far.

If the price increases, say it doubles during the time period we're looking at. If you don't wait after your initial purchase and you take the loans immediately, you'll owe 1BTC's worth of USD to Nexo at 6%. So assuming BTC doubles immediately thereafter, you're out the money you paid for your initial BTC AND the money borrowed against it to purchase a second BTC. So, 2 BTC in the red column. If you sell your 2 BTC at double the starting price, you will receive 4x the value you started with in return, but you'll have to pay back 1BTC worth to Nexo, plus interest.

So, you started with 1, and you end up sitting +200% instead of just +100% when the market price of BTC doubles. The downside is that you've doubled your exposure to risk as well as to reward, so when BTC doesn't double, you're on the hook to a troubling degree: if it goes down 50%, you own 2 BTC but you have to sell them both to pay off the loan that bought the second one! You've lost your initial investment!

Ideally, we watch the market and try to mitigate any downward potential by selling, but any loss—10% or 40%--will be doubled by our insidious risk multiplication machine here.

Conclusions

That's about the best I can do to really try to understand the appeal of DeFi, even as it wanes just a bit in terms of volume. The simplest explanation for the DeFi boom is that, in a bull market, people really like measures such as these which allow us to overleverage and gain that added exposure because the trendlines all point up and we would rather make more money than less money. The novelty of the technology piles onto that possibility of benefit to create excitement and, suddenly, a thriving market exists where none was before.

Still, none of these measures are without a certain element of risk. I never give anyone any advice in these essays, but if you choose to take something away from this article please choose this: tread lightly in DeFi. Applies to markets in general, in fact. Nobody really knows what's about to happen, and that uncertainty is a thing full of both danger and opportunity.

A Coordination of Nations, Central Bank Digital Currencies Gain Steam

12

Written by Owen Fernau

Macro investor Raoul Pal recently posted a Youtube video outlining what he views as the upcoming changes to the world-monetary system. Central bank digital currencies, Pal posited on Twitter, will be

"... the biggest overhaul of the financial system since Bretton Woods."

For those unfamiliar with the Bretton Woods agreement, the name refers to an international agreement that came on the heels of World War Two. The system entailed a guarantee that 35 US dollars would be redeemable for an ounce of gold. It also pegged all currencies of the involved countries at a fixed rate to one another and hence, to gold.

Bretton Woods also spawned the International Monetary Foundation, the World Bank to-be, and established the dollar's place as the de facto world reserve currency. It represents the most significant coordination of nations' financial systems of all time.

Until now, perhaps.

With the Bank of England's governor cryptically suggesting that a new digital currency would help the world move away from dependence on the dollar, and the IMF calling 2020 a "New Bretton Woods Moment" while holding seminars on central bank digital currencies, it's clear a major transition is in store for the world economy.

Let's see what central bank digital currencies (CBDCs) entail for the world.

Citizens will be able to send money instantly, across borders.

With CBDCs, people will have a direct link to other peoples' accounts across the world. Daily crypto users won't find this revolutionary, but for others who want to send money internationally, this will be a game-changer. International transfers often take days, usually with seemingly unnecessary transaction fees tacked on.

Central Banks will be able to send cash to citizens directly, thereby increasing the money supply through direct payments, rather than banks.

Central bank linked accounts will give the monetary organization a new tool with which to expand the money supply. In giving citizens money directly, Central Banks would be able to forgo the controversial quantitative easing tactics, whereby private banks' assets are bought, with the hope that those banks will then lend out the additional cash. This did not happen in wake of 2008.

CBDCs would also increase transfer payments' efficiency. As seen in the US with the COVID-induced stimulus payments, it took many months to get people their $1200 check. With CBDCs these transfer payments can happen instantaneously.

In the long term too, central bank linked accounts could allow for the simple distribution of universal basic income, a paradigm gaining steam as global growth slows.

With CBDCs Central Banks Can Also Customize Interest Rates.

This means that interest rates could change on a person to person basis based on big data algorithms' output. Students could be given higher interest rates to promote saving, while wealthier, older savers could be given lower ones in order to unlock their capital.

Custom interest rates mean behavioral economics will truly be integrated into federal systems. It won't be too dissimilar from the way Facebook and other tech giants use data in order to incentivize certain behaviors, in their case, monetization.

Those worried about the implications of central banks pulling economic puppet strings will vie for CBDGs with a queryable blockchain so citizens understand the algorithms dictating the interest rates.

It's not cut and dry good versus evil however. Targeted higher interest rates can help lift society's unluckiest people into a stable livelihood.

With All Transactions Monitored, Taxation Could Become Much Easier.

CBDCs could mean the end of the IRS. With every transaction on the blockchain, central banks will be able to tax at any point in the money flow they wish to. Another dramatic shift, with its own sets pros and cons.

So What Happens with Stablecoins?

They may be pushed out by CBDCs. Perhaps more likely is that Central Banks will require stablecoins to be fully backed by federal reserves. Time will tell.

What About Bitcoin?

With interest rates continuing to drop towards zero, the central bank is running out of options to stimulate growth. The economy is stalling. Some say it has never really recovered from 2008. Federal Governments may need to pay their way out of the slow economy through massive stimulus. That probably means inflation which in turn means that Bitcoin's price will rise as people search for a strong store of value amidst the chaos of a deflating currency.

Raoul Pal is on record saying he thinks Bitcoin will hit 1 million dollars by 2025. How much a million will buy though in five years, what with possible inflation, is another question

How Could Guerrilla Marketing & Blockchain Promote Culture Within Our Cities?

13

Written by Alex Cuevas

Cities include cultural highlights which most tourists and even local residents don't notice, overlooking an endless alternative list not visible on traditional city maps, in order to solve it, guerrilla marketing and blockchain would be a valuable tandem.

Guerrilla marketing is usually linked to brand advertising in public spaces, but this fact does not exclude other groups such as citizens, state entities, or collectives, which can also carry out actions with a specific goal in mind, which can be to foster culture.

On the other hand, blockchain technology involves not only accounting registers, but also a tool for decentralizing data, making it easier for citizens to create instruments to respond to their demand for cultural interaction, despite the inaction of local institutions.

Historical Data on Street Names

Sometimes street names are linked to specific events, historical places, or influential people, so perhaps it would be a helpful insight to include biographies of historical personalities or short local events stories.

Certainly, many countries use different methods for identifying their streets, such as numbers, in which they would have to create different strategies. In the case of my city, it is full of streets that mention philosophers. Why not include accessible biographies with a QR code?

There are some exciting decentralized platforms that could be exploited for this purpose, for example, everipedia.org works in a decentralized way, encouraging user participation by sharing relevant content.

In our case we can use this platform in order to share cultural information regarding our neighborhoods, for example, people can find out local history by walking around the street just photographing QR codes on particular street names.

I really find it hard to accept why these very simple actions are not yet widely implemented by the local administration, so perhaps we citizens could start placing QR code stickers on street names, or would this be perceived as vandalism?

Quizzes & Riddles Around Cultural Events

In my opinion, it would be great that cultural promoters give away some tickets around museums, theatres or music halls, albeit in a participative format, rewarding people who really enjoy culture.

An idea that could possibly reach this goal could be to create some secret quizzes in previous days to a specific event. For example, when there is an art show at a museum, a mini-competition can be organized where the first participants who correctly answer the question will unlock a wallet containing the precise token amount for a free ticket.

Decentralized Routes and Calendars

Blockchain technology can contribute to building decentralized itineraries according to specific interests, both residents and tourists, thus broadening traditional guiding alternatives.

As with street names, I believe Everipedia can help us in this case, but in addition, it is possible to develop a local dapp with a decentralized local cultural events schedule.

Local Artist Contests

Each neighborhood is home to amazing artists who can help increase culture on the streets through their creative works. In my opinion, temporary street exhibitions could be organized in order to identify and select better artworks to bring into the district.

In order to ensure a transparent contest, blockchain technology, and temporary tokens can be included among the voting participants, so winners might be rewarded as additionally listed on future neighborhood art projects.

Historical A.R Maps

Imagine the mere act of pointing your camera at a road to reveal how it was this specific street 100 years ago, or even how it is planned to look in the next 10 years.

Why not? Involving other technologies like augmented reality in addition to blockchain, we could take photographs within our cities, checking their evolution and registering significant variations.

Little Free Libraries

One of the best feelings for an enthusiastic reader is to share an inspiring book with a fellow. Our cities have culturally attractive spaces, however, there are places inside cities where it is even more interesting to carry out this concept, wherever accessibility to culture is unattainable for most people, who do not have a library in their district.

Using blockchain technology we could keep a comprehensive record regarding books and writers, and in addition, about the involved users. In this way, we could reward most active users with "cultural tokens" aimed at buying books from the writers who shared them, thus also contributing to buying new books from the writers.

Archaeological Data Records

Sometimes we walk around unaware that just below our feet there lived fascinating ancient civilizations, which are intrinsically linked with our remote and more recent ancestors.

Including archaeological excavation registers using blockchain can greatly facilitate the monitoring of proper conservation and maintenance, and also provide quality info.

Now it's your turn! Which measures would you choose to adopt blockchain in the cultural scene? Thanks for your support.

The Story Behind the EOS Detroit Proxy Prospectus art!

14

Written by Lars Kommienezuspadt

I met Robrigo, the CEO of EOS Detroit, serendipitously at a blockchain meetup in Minneapolis years ago. Steemit was my red pill, and through fellow Cryptowriter author and mutual friend, Mark Bailey, I began blogging and earning crypto through the platform. Mark invited me to this get-together, and to be honest like Morpheus warned, I wanted to see how far down the rabbit hole went. I was open to anything. New people, new ideas, new experiences. Along with a few other notables, Robrigo and I met that night.

There was a persistent obstacle in the way of my hiring, and that was a relentless bear market. One I'm sure I don't need to remind us of. You see, we're in this transitional phase right now. Blockchain trends are fueled by buzzwords. ICO's and IEO's. Usability and interoperability. Web 3.0. DeFi. NFT's in the latest and growing the fastest right now, but I assure you, when I was pitching the strategy of leveraging art to shape EOSIO culture, it really wasn't. We saw the rise and fall of Cryptokitties regarding the potential for NFT's, but aside from that and similar projects, art was a second class citizen to length, boring Medium posts touting transaction speeds and white papers.

I envisioned a future where EOS Detroit would use my illustration, art, and storytelling to infiltrate not only the minds of those already initiated in this revolutionary ecosystem but imaginations of the regular people who have no conceptual understanding of what a blockchain is. The masses who drop off in the first 15 seconds of a TPS report or governance model theory. Art would be our Trojan horse, and for all the great work EOS Detroit had done since the go/no go in June of 2018, I wanted to convey those achievements in ridiculous, fantasy fueled comics and graphic narratives.

Eventually, we emerged from the bear market and Robrigo was able to offer me a position giving the company a fully-fledged, capable art department. One of my first tasks in this role was to finish our Q1 Proxy Prospectus. This is a biannual publication that presents our community with company and community updates, partnerships, achievements, and transparency information. We released our first Proxy Prospectus under my position as Art Direction in May of 2020. I'm very pleased to bring you our next iteration in Q2/Q3 Volume IV.

I'm really proud of this release, and of course, I do encourage everyone to read through it. More than just an opportunity to shill the release, though, I wanted to use my platform here on the

Cryptowriter Voice community to share insights from the pages of the Proxy Prospectus and how I came to these illustrations; starting with the provocative cover, which was intentional. Some relevant geography is my home, Minneapolis, Minnesota. After the killing of George Floyd by a Minneapolis police officer, our city became the epicenter for a social revolution.

Protests began in nearly every major US city, but a few, in particular, I couldn't help but notice a remarkable development. Minneapolis, Portland, and Seattle, specifically. You saw them portrayed in the news as "autonomous zones", but I saw them [first hand, as I visited George Floyd Square at 38th and Chicago in Minneapolis] as real-world, genuine DAC's, or decentralized autonomous communities. My familiarity with blockchain governance and crypto-anarchism wouldn't allow me to ignore the beauty of that space for the time it existed.

As far as I know, the space in Minneapolis is still that; a decentralized community safe zone, where needs are met by the community, including food, clothing, shelter, and healing. Free Capitol Hill in Seattle also emerged as an autonomous zone, and it was these developments across the country that lead me to the cover for the Q2/Q3 Proxy Prospectus. As much as it's a statement on where we are as a society, it's an acknowledgment of decentralization being effectively implemented outside of the blockchain. For me, a revelation that anarchy is less about *no one in charge*, and more about *all of us in charge.*

I'll be honest, at 40 years old, my romanticism of anarchy from the time I was an angsty punk in the '90s has surely faded. Over the last five years, though, I've been actively present at flashpoints pivotal in what society deems as "too far" by our government. Philando Castile was murdered two miles from my home. I was called to that injustice in protest at the Governor's mansion. Standing Rock, which was probably the most eye-opening experience of my life in regards to abuses of power by our elected leaders. Our phones, the only tool we had to tell the world what was happening to Lakota elders, being shut off by Federal police with EMP's [...which I'll remind everyone was during 2015, during the Obama administration].

This was a traumatic summer, especially here in Minneapolis. I grew up on the east coast, in a German family. "Trauma" isn't something often observed. "It is what it is.", they'd say. I will tell you, after George Floyd's death and the endless images of violence and death, I'll admit there is an undeniable, immeasurable degree of PTSD with us. Just a few days ago, I heard a helicopter fly above my house, and I immediately felt my heart begin to pound. Checking on my son and instinctively running to lock the doors. We're on a knife's edge, and like so many in this space, I see blockchain and the philosophies within the technology, as a hopeful solution to many of our complex social challenges.

The rest of the Proxy Prospectus for this release focuses on the achievements of EOS Detroit. We've had a remarkably productive year, despite the pandemic, social unrest, and economic hardship. We've grown, adding diverse, talented team members such as Cryptoblood as our new CMO. We've made big investments in NFT strategies early on and those efforts are bearing fruit. We've partnered with Cryptowriter to bring Finney, the inquisitive Android to the face of the company. There's a lot to be excited about and I feel lucky to be positioned to do so much art in such a revolutionary space.

How Would Blockchain Improve Accessibility For Blind People?

15

Written by Alex Cuevas

We all hope that some areas such as technology and medicine will move ahead swiftly in giving full autonomy to blind people, and to ensure this, blockchain can play an additional role, especially concerning certification and transparency.

Thus, it is essential to consider blind people's perspectives, otherwise, it would be senseless to ignore those who are directly concerned when trying to design accessibility strategies.

It is worth mentioning a brief background that emerged in the crypto ecosystem in this context. Six years ago, a Reddit user (gamondbozia) posted on r/bitcoin about how hard it was to manage bitcoin wallets, feeling excluded in the community, and rightly so, as no bitcoin wallets had considered blind users like himself, who were not even able to send or receive transactions without his sister's help.

Discussion subsequently led to some improvements in the crypto ecosystem concerning blind inclusion, for example, IcyWallet, the first 100% accessible bitcoin wallet for blind users, or a great braille version of bitcoin whitepaper uploaded by neatnik at GitHub.

Several positive developments have already occurred in the crypto ecosystem for improving blind accessibility, but today, in contrast, we will explore some of the potential gains of using blockchain technology in other areas, from the perspective of blind people who are not involved with crypto.

So, how can blockchain support blind people in other real-life situations?

Web Accessibility Certificates

Theoretically, most countries must assure all their websites are accessible for people with particular health conditions, including blind people. Unfortunately, this does not usually apply, becoming a real issue for blind citizens who are unable to access online content offered by public institutions.

A few years ago it wasn´t common to include the necessary code to support blind tools such as screen readers, but nowadays, big companies are starting to see more relevance to the Accessibility Software Engineer and I think public institutions should aim in the same direction.

Accessible Thinking Contests

Commenting inside a blockchain transaction is not a novelty, so it wouldn't be difficult to create a token that serves citizens to contribute funding to accessibility research and at the same time allows them to leave innovative ideas, which are then rewarded when used to implement it in the community.

Decentralized Visual Descriptions

Imagine for a moment how difficult it would be for yourself just walking along streets with absolutely no vision. Despite the familiarity of the area, common spaces can change completely every day, so a visual description could assist blind pedestrians by warning them of possible alterations, for example, advising if a pavement is under construction at the moment, if a traffic sign has changed, or simply indicating time and weather data.

In this way, public institutions can create apps where both citizens and blind organizations with other professional entities can include descriptions, which can be validated in a public or private blockchain, accessible to everybody or some groups, in order to provide real-life environmental information for blind or low vision individuals.

Incentives For Accessible Companies

Opposite to public websites, businesses are not required to provide accessible web content, so as it is not compulsory, I think that an incentives plan which rewards companies that are accessible is a great starting point for raising people's minds from the business environment.

For example, incentives could be developed at a 1-10 scale, being 10 the maximum score, and data might also be collected as tokens on a blockchain, so citizens are able to check which companies are becoming more accessible to the online world in real-time. In this way, blind people know exactly which companies' websites are currently accessible on the Internet.

If you would like to express any ideas on how to improve accessibility, all suggestions are welcome. Feel free to express your opinion in the comments section. Thanks!

Leveraging NFTs on Blockchain-Based Social Media

16

Written by Sean Ballent

If you haven't heard, Non-Fungible Tokens (NFTs) are currently the hottest thing in crypto and are poised to take over the world. Once you fall down the NFT rabbit hole you walk the stairs of an M. C. Escher printing. A peregrination of possibilities, NFTs will shape the future of art, collectibles, gaming, and beyond. They provide a synergistic leveraging opportunity for blockchain-based social media platforms to help grow communities and add utility to their native tokens.

NFTs as Badges

NFTs in the form of badges can be used for user-customization, rewarding, and for content filtering. Platforms could empower users to show a little personality by allowing them to select and display badges in the form of NFTs across the top of their profile/posts. Imagine when signing up for a platform, a library of badges, similar to a library of emojis, where you could select up to 3 or 4 to identify yourself with. These identifier badges could also be provided by the platform to identify genesis users, community moderators, and influencers across the platform.

Reward mechanisms can help drive content and behavior across social media platforms. Blockchain-based platforms can utilize NFTs to reward users for things like exceptional content and dedication. NFT badges can be used to reward milestones, engagement, and top content creators.

To take it a step further, imagine users being able to filter content by a badge. Maybe you'd like to see all the content by "top content creators" represented by the "photography" badge. Or maybe you'd want to view only content by those displaying the badge of a particular country.

Adding Token Utility Through NFT Marketplaces

Blockchain-based social media platforms have a unique advantage over traditional social media platforms in that they can build economies around their native tokens. Over the last year, crypto has seen an incredible NFT boom relating to digital art NFTs, collectibles, and the marketplaces where they are traded. Blockchain-based social media platforms should build NFT marketplaces into their architecture, allowing users to trade digital art and goods using the platform's native token.

Going back to the badge idea, imagine having a baked-in marketplace where any user of the platform could design and sell customized badges to other users of the platform. Any user who wanted to could participate. Once purchased, these badges could be displayed, gifted, or stored to be traded at a later date.

What works for badges could be used for art. Imagine if the platform's Digital Art community had its own marketplace and gallery where community members could display and trade their amazing art.

Imagine a Music community marketplace where digital downloads could be sold for tokens in the form of NFTs, allowing artists to get paid for their content. The possibilities are virtually endless.

Blockchain-based social media and NFTs are a perfect marriage. Not only can NFTs be leveraged to grow communities, but they can add utility to the native token. What are other ways you can see the integration of NFTs and social media?

Token Staking: Cost-Zeroing A Position

17

Written by Thomas Daniel

Thomas Wolf recently penned an article about the omnipresent question of how staking works for those who do it. I feel that Wolf did an excellent job of explaining the downside but perhaps was a bit too focused on Tezos (XTZ) to really comprehend the major upside to token staking. This article will look at BAND Protocol (BAND) and Cosmos (ATOM) as perennial staking solutions that can allow a trader to achieve a bit of safety at the end of a bull market.

Flipping coins will always be the fastest (and riskiest) way to invest, but staking in the Cosmos ecosystem comes with a 21-day unbonding period that provides users a surprising advantage. The point is, when you stake with Cosmos or with BAND or even with certain other tokens such as Zilliqa (10 days instead of 21, but @

25% APY in ZIL), you're locking up tokens that you might otherwise decide to sell. Volatility in the market decreases when many people exercise this option, and price stability remains far greater for ATOM than it does for XTZ as a result.

My strategy thus far has been to look at these tokens and make small, cautious investments into the ones which appear to have the best return. Though I'm not an expert, a few quick googles can reveal the market's current appraisal of the security of a new blockchain as well as its potential for adoption.

The Goal of Cost-Zeroing

The end result we're all looking for is a HODL that will pay many multiples of the initial investment. If we invest our money, we want a return on that investment. Simple as that. For the past several months, I've struggled to pay bills due to my overzealous embrace of token staking, but there have been some surprising benefits. I earn a passive income of a few hundred dollars a month in a variety of cryptocurrencies, and my endgame is to HODL until these tokens are worth enough to pay my expenses as I sell them.

I want to stress that this strategy is one that we can consider fairly extreme, and I'm applying it in a variety of cases to tokens that provide large returns to users not for the returns themselves, but in hopes that the market for these tokens does well and provides me a direct return on my investment value.

Thus far, my record in this area is mixed. I bought BAND before the September crash and then proceeded to buy all the dips, bringing my cost average down to about $8. My hope is that new all-time highs are on the way in response to BAND's interoperability and low cost in the oracle market, combined with the security of DBFT (decentralized byzantine fault tolerance is a popular Proof of Stake model employed by the Cosmos ecosystem). I'm staked at 17%, adding BAND tokens to my bag for free as I wait.

My Cosmos bet was far better timed, back in late June, and I was able to enter the market for under $3. Thanks to the Proof of Stake model, I have been selling tokens ever since. Some I've sold for north of $7, some I've sold for around $5. But I've paid my living expenses out of my Cosmos stake and I'm happy to report that my cost basis for the remaining position is rapidly approaching zero.

As You Pull Your Money Back Out, Diversify!

I've decided to hold all of the Cosmos that will accumulate after cost zero and set my sights on other tokens, including BAND and Zilliqa (ZIL). Zilliqa is by far the sketchier of the two, but it is more liquid than Cosmos staking due to a shorter, 10-day unbonding period. Plus, I was able to enter during a very low point for the market, so the hope is certainly that the position will be healthy whether or not the Zilliqa blockchain increases its market cap by multiples in the near future.

As I nurture my small ZIL and BAND positions via my Atomic Wallet, where I also hold AWC that's down from where I bought it but almost back to even due to staking rewards payouts every week, my Bitcoin and Ethereum positions are outperforming these mainstays of mine by a bit. I am not worried by this, and in fact, am happy to have a source of profit to keep myself afloat as I patiently await the next bull run.

It's important to diversify positions when investing, not only because this minimizes risk, but also because it can increase exposure to surprisingly profitable assets such as Civic (CVC), which is making a real run at it for reasons I don't claim to fully understand. Still, my small CVC bag is worth over double what I paid for it a few months ago and I'm glad I bought and held it. Nonetheless, when it comes time to cash in CVC, the operation will be a simple flip for which I will need to spend a bit of ETH. I am reminded of my charming staking rewards and the various strategies I can use to dispose of the gains they create.

Cost-Zeroing

In order to cost-zero a position, there are a few things you'll need to do.

- Buy some sort of token with staking rewards and easy reward tracking.
- Ensure that the coin you're getting into has a future; i.e., that the price has room to rise.
- HODL and accumulate rewards.
- Sell off some of your staked assets when the price rises.
- If the price rises enough for you to keep ½ of your initial investment staked while reclaiming your initial investment, *do it.*
- Create another staking farm by placing your initial investment into a new asset, use it to pay bills, or wait for a low price and buy back into the initial asset if things are going well for an additional profit.

Why Cost-Zeroing is a Smart Long-Term HODL Strategy

We want to maximize our efforts, and that means the research we do as well as the gut reads we all do on the market from time to time. Some traders like to split their portfolios between traditional HODLs and day-trading and/or flipping plays, but for me, the day-trading is a bit too tedious unless there's a really strong news article or other sudden and significant upward pressure that's easily discernible and likely to be seen that way by a large majority of traders.

With PoS crypto tokens, the HODL strategy is augmented by the ability to passively earn a return on your investment provided the price of the token does not plummet too dramatically. ATOM in particular manages to guarantee this by ensuring that a majority of tokens are locked up at any given moment. If the price dips too low, traders buy up the liquidity in the market and stake the tokens to wait for it to rise. If the price rises too quickly, traders may unstake some portion of their staked assets but most see ATOM as a long-term play and thus aren't too inclined to unstake on the news of the day.

Cost-zeroing is the longest of long-term HODL options out there today. With ATOM, your investment doubling time, assuming a steady price, is 7.2 years today assuming a return of 10%. With BAND, your 72-rule doubling time assuming you compound the proceeds of your investment is 4.2 years assuming the 17% return you can find in the Atomic Wallet.

By staking and unstaking your assets according to projections of where the market should be, you can ensure your ability to continually sell high and buy low while also generating a reasonably large return even in the lulls between bull runs.

- Price drops?

Average down.

- Price bounces?

Sell whatever you have unstaked and look for a new opportunity.

These token sales at higher prices than point of entry over time allow us to withdraw our initial investment over time, averaging the cost basis of the position down to zero. Though it can take months or years or never happen if you choose the wrong token, this strategy allows for positions played wisely to stretch and cover the entire market. As most of what's going on in crypto right now is incredibly bullish, that seems like one of the smartest plays an investor could make today.

In my situation, liquidity is one of the most important considerations I routinely make errors in. I've devised a strategy that incorporates BTC and ETH investments in a Nexo wallet, and I've learned to sell off the staking proceeds my positions generate during bull runs to help myself stay afloat despite limited income due to the pandemic. Managed wisely, a strong position in an appreciating staking asset is one of the best things that can happen for you in crypto trading today.

From Silk Road to Digital Dollars

18

Written by Mark Bailey

When I first started looking into crypto, Bitcoin was the only game in town. As that began to change, I became increasingly excited about cryptocurrency in general. The idea that any community could create its own digital money outside the parameter of the banking industry seemed particularly compelling. My excitement was, however, tempered by a fear that the allied forces of business and government might treat the whole cryptocurrency space as illegitimate. This fear disappeared when the US government shut down Silk Road, the illicit online marketplace, in 2013.

Silk Road Forfeiture

The US Justice Department initially seized 144,336 Bitcoins from Silk Road. The government then sold this Bitcoin for approximately $48 million through a series of auctions in 2014 and 2015. To my way of thinking, these auctions proved that crypto was a legitimate currency. I bought my first Bitcoin in 2014.

Recently, the US government seized another 69,000 Bitcoins, valued at approximately $1 billion, from an individual who apparently stole these tokens from Silk Road between 2012 and 2013. It's not yet clear if these seized assets are to be auctioned off like the rest of Silk Road's treasure trove was. But the fact that the US government dedicated resources to seizing them further bolsters the legitimacy of the crypto space in general.

This is not to suggest that such forfeiture procedures are, themselves, legitimate. Legally, they seem dubious. Police have a long history of abusing asset forfeiture laws and the Supreme Court occasionally rules against police in forfeiture cases. But the act of seizing property, itself, legitimizes the value of the property. The government wouldn't seize cryptocurrency without compelling reasons to regard it as valuable.

Digital Dollars

In the seven years following Silk Road, the crypto space became fully legitimate. Last spring, legislation was proposed in Congress to create digital dollars. One such piece of legislation, the Automatic BOOST to Communities Act (ABC Act), defined 'digital dollars' as follows:

The term 'digital dollars' means dollar balances consisting of digital ledger entries recorded as liabilities in the accounts of any Federal reserve bank and digital coins or currency instruments issued by the United States Treasury as legal tender, and held as bearer instruments in any digital dollar cash wallet approved by the United States Treasury.

The primary objective of the ABC Act was to give away $1 trillion in pandemic relief to individuals, which proved to be a political nonstarter. But future legislation that would create digital dollars may fare better. *Forbes* recently ran a piece about 'central bank digital currencies' competing with cryptocurrencies like Bitcoin. And the *Wall Street Journal* titled an article, 'The Coming Currency War: Digital Money vs. the Dollar'. This media coverage suggests that digital dollars are coming eventually.

The Future

Seven years seems like a very short time to go from Silk Road to digital dollars. It's a timeframe that puts things in perspective. For now, central bank digital currencies like digital dollars are only theoretical. But seven years from now, digital dollars might be ubiquitous. If digital dollars do become ubiquitous, I hope they play nice with crypto.

Specifically, I hope digital dollars are created in such a way as to make them easily tradable on all the major crypto exchanges. I also hope that decentralized exchanges become as useful as centralized exchanges, so these digital dollars can be swapped between parties without intermediaries. Even if these hopes turn out to be unrealistic, the next seven years will obviously bring major changes to the crypto space. I am curious what you think the most important changes might be.

Will digital dollars become reality? Will a coin that hasn't been invented yet come to dominate the market? Will Bitshares make a comeback? Will the IRS start accepting payments denominated in crypto?

My experience as a Tezos crowd sale buyer suggests that crypto-related things can take much longer than anticipated. The things we expect to see in two years might end up taking seven years to manifest. I expect to see crypto tokens by Facebook and Google fully rolled out in two years. What changes do you expect to see in this space in two years? In seven?

Why Do Central Banks Trust Ethereum's ConsenSys?

19

Written by Dalmas Ngetich

Central Banks and governments have had an evolving relationship with cryptocurrencies. In the early years of Bitcoin, crypto was met with a sense of dismay and largely ignored. However, the more popular Bitcoin and other cryptocurrencies got, the more difficult it became to ignore them.

Fast forward to 2020 and some central banks are exploring ways to channel this historic innovation in their favor. Central Bank Digital Currencies (CBDCs) are a hot talking point. These are essentially digital forms of central bank money, which will likely become legal tender. For crypto, this could be a watershed moment, even though it undermines decentralization.

But that's just one aspect of this. The overarching theme is that Central banks now look at crypto as a legitimate form of finance, only that they don't relinquish control.

Ethereum ConsenSys to Partner with Some Central Banks

Ethereum developer, ConsenSys, is set to partner with a number of central banks in researching the viability of CBDCs. It is an American software company that helped in the development of decentralized software services and applications that operate on the Ethereum blockchain. Besides that, it offers malleable blockchain solutions that can underpin the rollout of central bank-backed digital currencies (CBDCs).

With its wealth of expertise in building developer tools, enterprise applications, and other aspects of blockchain development, this company has what it takes to contribute to CBDCs.

The Reserve Bank of Australia announced that it would partner with ConsenSys and a few other entities to explore the possibilities of wholesale use of CBDCs and Decentralized Ledger Technologies (DLTs). The primary focus is to develop a Proof-of-Concept (POC) for the issuance of a tokenized form of CBDC. Research on the viability of such projects is essential to observe both the viability and challenges that come with such a bold implementation.

Additionally, ConsenSys will partner with the Central Bank of Thailand to explore the potential of blockchain finance. For now, Thailand has been working on a CBDC called Inthanon. ConsenSys could help with the creation of an Ethereum alternative.

At the risk of understating the significance of these developments, these are exciting times for cryptocurrency as a whole. ConsenSys gained credibility with its great work in the development of Ethereum. This credibility means that the company will have a starring role in the creation of CBDCs worldwide.

Benefit of CBDCs on the Blockchain

CBDCs can be great for making the monetary system faster and low-cost. This would make the financial system more efficient.

The basic element of a CBDC is that it is backed by the central bank and will have the legitimacy of legal tender. Cryptocurrency has not had the adoption levels of major national currencies because it has operated largely without the support of governments.

CBDCs are a game-changer in this regard. They are central bank-issued and controlled and can be a boon for retail and wholesale settlements. For retail, CBDCs will have benefits like Increased availability, improved digital information, and enhance monetary policy because the government can spread benefits to individuals more directly.

Ethereum developers like ConsenSys will play a vital role in the launch of these CBDCs, and rightfully so. Many Ethereum startups have rich data and workable models on rapid transaction digital payments. These implementations can be a valuable resource in unrolling CBDCs. Ethereum is a production-ready blockchain and can handle the scalability and privacy needs of CBDCs quite well.

The advantages of Ethereum as a launching medium are wide-ranging. For instance, the platform's programmability allows central banks to code CBDC rules into the minting protocol for compliance purposes.

Additionally, the Ethereum network is highly reliable and resilient. It has never stopped functioning since launch. With reliability, there is data availability, a feature that is key to the success of on-demand projects.

Furthermore, from Ethereum--one of the oldest blockchains and the pioneer smart contracting platform, users can wholly trust the system for their innovation and advocacy for complete decentralization and independence.

The clamor for better systems has been around for a while but nothing has ever challenged the monopolistic SWIFT international interbank settlement layer like cryptocurrencies and emerging Fintech platforms.

A wholesale implementation of CBDCs will mean a new infrastructure for interbank settlements, reduction of counterparty risk, participation in digital asset markets, and general competitiveness in the digital economy.

The existing financial settlement layers are costly. Moreover, CBDCs bode well with the general digitization of finance over the past few decades. CBDCs can eliminate the inefficiencies of the existing central bank infrastructure. In retail, CBDCs can streamline reconciliation and make transactions faster and cheaper.

Regardless, the logistics of nationwide implementation of a CBDC are immense. Monetary control is a delicate concept and it would be understandable that Central Banks are going to take the research and testing steps of a potential CBDC very seriously.

Focus on What the Fed Will Do

Given the size of the US economy and its extensive influence in the financial world, there is perhaps no entity that could do more for CBDC legitimacy than the Federal Reserve.

For instance, the recent monetary stimulus packages to every American family would be much more efficient with CBDCs rather than using checks that take too long to reach intended consumers.

The Fed has reiterated its commitment to inject liquidity into the American economy. A CBDC could be a direct way to transfer money to users and open the floodgates of liquidity. Besides, CBDCs greatly enhance the velocity of money within the economy. There is obviously the risk of triggering inflation with this strategy, but a lot of things that the Fed has done have already done that.

Therefore, it may only be a matter of time before the Fed sees the writing on the wall. CBDCs have many advantages over traditional money. Research with insight from experienced players like ConsenSys will make Central Banks around the world realize these possibilities soon

Why Gold Bugs Still Don't Dig Bitcoin

20

Written by Jason Deane

I was lazily perusing some of my favorite charts this morning (which, as an analyst, is not uncommon) and happened across a gold price chart from the rather appropriately named *goldprice.org* website.

Now, it's true that most of my work is on Bitcoin, but there are some significant similarities between the flagship cryptocurrency and the yellow metal, so this is a chart I refer to from time to time (in conjunction with others) to get a broad view of what's happening at a top-level.

What was striking about this morning's chart was just how significant the recent pullback had been at a time when almost all other assets had risen. In fact, it took the price back to the same level it was at on July 22nd—a date that now seems like a point in ancient history when you consider what has happened since.

Over the exact same period, the Bitcoin price has risen by approximately 58.47% to reach today's dollar value of approximately $15,400 at the time of writing.

What struck me is how unlikely that scenario would have been any time prior to this date and how counter-intuitive this outcome would have been if you'd tried to work it out in advance.

After all, apart from specific market sectors—especially tech—this has been arguably one of the biggest times of uncertainty in recent history. Covid-19, US elections, unprecedented economic stimulus programs—under the "old" rule book, you'd just buy gold and wait for it to all blow over.

But the world is different now. We're writing new rules, we're taking new risks and we're trying to make sense of a financial system that is under strain from all angles. In my view, nowhere is clearer of this happening than in this simple example.

However, as I went to click away and carry on with my day, I had another thought. Would hard-core extreme gold maximalists, the most famous and charismatic of which is Peter Schiff, see it the same way? Would they now consider it at least *possible* that value could be stored through a new, entirely "artificial" asset like Bitcoin?

The Gold Bug Blindspot

Peter is very active on social media and scrolling through his Twitter feed reveals no change instance in terms of Bitcoin. His current comments are visibly anti-Bitcoin, not just as a store of value, but even about having an existence at all.

This is not uncommon among what I refer to as the "Old Guard" of the economic world, and it's not just Peter.

Nouriel Roubini and Warren Buffet are other famous big hitters who also despise the coin with a passion, yet interestingly agree with the assessments of the underlying macroeconomic picture.

So, the strange situation that we find ourselves in is that we all agree that fiat is bad, the global economy is in trouble and it's up to the individual, company, or sovereign state to protect their wealth. They just see it that route to salvation as gold (or other assets) and not Bitcoin.

But these are very smart people—why would they not consider Bitcoin an option *at all*?

We Can Change Your Mind ... I Think

Listening to these people speak in general interviews gives us some clues, and more than one cryptocurrency advocate has tried to change the mind of each of them without success.

Sometimes we only get a summary of those encounters, such as Justin Sun's famous (and expensive) dinner with Warren Buffet, but sometimes you get to hear the whole argument played out.

Peter Schiff has regularly appeared on podcasts and shows, and I have listened to many of them, but my favorite is actually by fellow Brit Peter McCormack. This is a well-balanced discussion with valid points and solid rebuttals on both sides and it makes compelling listening.

You can read the transcript or listen to the entire podcast here.

There's no doubt that Peter (Schiff) didn't change his mind through the course of the podcast, but it does give some extra clues about where his thinking is at.

Bottom line? This is all about what I call the "wet paint" scenario.

Wet Paint and Bars of Bitcoin

We humans are very odd in certain ways, and often predictable. Consider, for example, the following situation:

Let's say you and I, following Covid-19 guidelines, got to meet in a park, something that is still allowed in the UK at the moment. As we go to sit down on a nearby park bench, we notice a sign that says "wet paint".

What's the first thing we'd do?

I can almost guarantee we'll touch the bench to see if the paint is really wet, even though there is a sign that quite clearly says it is. Not only that but even if the paint clearly possessed that "just painted" look, we'd still touch it, just to make sure.

There are all sorts of sensible reasons for this to do with checking that the sign hasn't been left after it was painted hours or days previously, or simply wondering just how wet it is, but mainly because there's a part of us that wants to physically confirm what we're being told. It doesn't change the fact that it is or isn't wet, but it does change our perception of that status.

Could it be it's as simple as that? Let's put it another way.

Imagine for a moment that Bitcoin had never been invented. Instead, in 2008 we had simply discovered a new mineral, with very similar qualities to gold, in small deposits around the world and learned to process and develop it.

Let's assume it has a specific use in the manufacture of, for example, a new generation of electric car batteries, but in reality, its value was derived from holding and speculation, in much the same way as gold. Mining industries sprung up all over the world searching for this new, scarce asset called *Bitcoinonium.*

In just over ten years, this commodity had become one of the most sought after assets on the planet, and bars of it were guarded in special facilities. Trading was now possible using the asset itself and derivatives such as futures and options. It was being adopted by all levels of high net worth individuals, funds, businesses, and even sovereign states that were considering holding it.

How would Peter and his colleagues react to this? Would they have been negative and seen it as a competitor to their precious gold, poo-pooing it at every opportunity?

Or would they have pivoted some of their business to secure some of this new asset and store it in their vaults using the same systems they already have?

My guess, especially based on comments in the podcast, is that they would have done the latter. Maybe not immediately, but at some point, it would have become difficult not to.

So, what's the difference?

Touchy Feely

The truth is that it is hard to get your head around Bitcoin's vast power and network, especially at first. It doesn't feel real or tangible. After all, if I can clone your hard drive and have a copy of all your music, surely I can copy your bitcoins too?

Well, you can't. Satoshi's genius was solving the "Digital Scarcity" problem, ensuring that only you have your bitcoin, until the point you irreversibly give them to someone else. It is the Holy Grail of value transfer which simply doesn't translate to a world of physical gold and requires a way of thinking that is incompatible with that mindset.

In short, Peter needs to "touch the paint" to check it's dry, he needs that absolute proof, and mathematics simply won't cut it. It has to be *physical.*

The Theory is a Theory

Of course, this is all a simplification. Peter cites many other reasons in this podcast and many other media appearances, as do the other opponents listed here.

All, however, have valid counter-arguments on a logical level, but this doesn't matter if the fundamental belief is that the paint is not wet when it is, especially if you, at least in your mind, have no way to check.

After all, we sometimes change our minds, but we rarely change our beliefs.

Warren Buffet is quite honest when he says he doesn't understand it and doesn't really want to (the man is 90 years old), and Nouriel simply won't even look at it enough to start understanding it properly. He still bases his opinion on the very early days of bitcoin when it was barely out of the sandbox.

That's like refusing to use the internet because when you dialed in with your 28.8kb modem in 1998, you thought the text-based websites were rubbish and it was really slow.

In any case, it doesn't matter. Bitcoin's march towards global adoption is now next to inevitable regardless of what gold bugs think and the truth is I believe both assets can co-exist and complement each other in the same way I believe that Bitcoin and whatever fiat-based system we'll be using can co-exist. It's just that Bitcoin will be better, more versatile, and more widely available than all of it.

Of course, this is all just theory made up of letters on a screen.

You can't physically touch the idea, so you'll just have to go with it.

Or not.

And therein lies the problem.

Tech, Distrust, and How the Blockchain Might Help

21

Written by Owen Fernau

"Tech is evil."

People instinctively hate "techies." I was at a Halloween party with some young people and was talking about working on the computer with a guy there and he hit me with the one-liner that "it's evil."

I knew what he was talking about.

A World Without Consequences…

The digital world is impermanent. You can turn it off with a button. You can't turn off physical items. An angry person online can be blocked and ignored. That same person in front of you must be handled.

Tech is a wealth vacuum because of the leverage that the Internet has. Everyone with a net connection is a potential customer. A digital product has unparalleled reach, so the winners of the Internet win big.

It's a very recent phenomenon. For almost all of human history we've dealt with intuitive physical constraints. When a person eats a loaf of bread, no one else can eat that loaf. A house can only hold so many people.

When a person uses an app, however, network effects actually increase the likelihood of usage for that same app. The more, the better.

The Internet made the reproduction of media and software, instant and free for companies.

The Conflict.

I have worked in bakeries, at cafes, and on construction sites, but I also know basic web development, CS, and have been involved in tech for the last few years.

It feels stupid not to work in tech because I can't unsee the leverage. But I sense that internal distrust of this ephemeral digital world which disappears with a click.

My uncle builds houses. When they're built, you can see them, touch them, and live in them.

On the other hand, I wrote words on a digital page as a freelancer in college. But the app I was writing for got shut down. My writing's gone. It's like it didn't happen.

This is where the blockchain comes in.

The Blockchain Brings Us Back to Reality.

This may be very old news to some, but the information stored on the blockchain entails orders of magnitude more durability relative to current information storage standards.

The blockchain is a permanent record. As long as the network runs, the information on it exists, and it can't be tampered with by any private entity. Conversely, Facebook owns its servers and hence can modify and censor the information on its platform. This has been discussed ad infinitum in the crypto world.

Less discussed is the psychological level.

Infusing the Digital World with Blockchain-enabled Scarcity Means that the World of Bits Will More Closely Resemble the Physical One.

Scarcity exists on the blockchain and in the physical realm, but not on the legacy Internet. Again, if I eat a loaf of bread you can't eat that loaf. Similarly, if I spend your Bitcoin, you can't spend your Bitcoin.

Pre-blockchain, the Internet has, among other things, gutted the arts. Images can be stolen, songs can be ripped, definitive ownership remains elusive. The blockchain changes this. Non-fungible-tokens means art can be attributed to a specific person.

This protocol generates ownership in the digital world, of both cryptocurrencies, and also unique, one-off items.

The Blockchain World Comes with Consequences.
If a person says horrible things on Twitter under an anonymous account, it's almost guaranteed nothing will happen to them. But if, say, your Ethereum address is linked to your social account, the things you say will be linked to your account forever (or as long as Ethereum exists).

Similarly, if a person has a bug in their smart contract, millions can be lost. This doesn't happen with normal software apps.

With the Blockchain, a Concrete Past Exists on the Net. Before, the Past Was Malleable.

The blockchain-based Internet, despite the futuristic aspects of global capital sloshing around crypto protocols, will more closely resemble the physical world, than the pre-blockchain Internet has thus far.

Moving Forward
It's difficult to avoid a delusively positive outlook on the blockchain space to which I've dedicated so much time. Everything written above has many caveats. Still, I'm optimistic that over the long term the blockchain will help the Internet feel more like the real world. Not a utopia, but a real-world, with consequences, scarcity, and people summoning the necessary deliberation for navigating such a world.

The stakes are extremely high for smart contract engineers writing code that will facilitate citizen's transactions. Similarly, the mechanical engineers designing public bridges must make absolutely no mistakes or assumptions.

These jobs are more consequential than using Javascript to get an animation right (no knock on designers, product design is as important as ever).

But the period of time between 1990 and 2020 may be an aberration in terms of the Internet's childish culture. The commercial Internet is 30 years old after all. It's about the time it, and we, and I, grow up.

As Confucious said, "At 30, I had planted my feet firmly upon the ground".

How Bitcoin's Power Consumption Is Good for the Planet

22

Written by Jason Deane

A Controversial Proposition That Might Just Hold Water

In early 2018, an analyst called Alex De Vries at Pricewaterhouse Coopers (PwC for short, thank goodness) produced a report entitled "Bitcoin's Growing Energy Problem" which seemed to provide a plausible calculation of the power consumption of the Bitcoin network.

It seemed viable, was generally well supported, and, rightly on wrongly, went on to become the basis on which many lasting assumptions about Bitcoin's power consumption were made. It was timely enough that I even included a detailed reference to it in one of the chapters of the book I was writing at the time called "How to Explain Bitcoin to Your Mum"

The chapter was called "Why Bitcoin WON'T work" which was a counterpoint to the preceding chapter entitled, surprisingly enough, "Why Bitcoin WILL work"

Some, however, questioned the validity of the underlying assumptions about the type of equipment used as a reference point to calculate power consumption, and some considered whether renewable energy had been given enough consideration overall.

While both points are far from resolved, the latter was revisited by the same analyst a year later when he produced another report entitled "Renewable Energy Will Not Solve Bitcoin's Sustainability Problem", the main conclusion of which is probably self-explanatory.

The Size of the Problem

De Vries had done the best he could with the data available, as had many others when trying to answer the question of just how much power Bitcoin consumes, but the actual number varies considerably depending on who you ask and at what point in Bitcoin's history you need an answer for—and with good reason.

The Bitcoin Energy Consumption Index has long since estimated total power consumption and, as of today, estimates a figure of 74 TWh (Terrawatt Hour), which is roughly equivalent to the entire power consumption of Venezuela, at least according to their latest official estimate which, oddly, was way back in 2015.

Even getting accurate energy consumption data by country is surprisingly hard and often the range of official estimates varies according to which source you refer to. No single source, in fact, appears to have a consistently up to date list of energy consumption levels, a point I find interesting in itself but is beyond the scope of this article.

Nevertheless, it is certainly *possible* that Bitcoin's current power consumption is more than Denmark, Hungary, Hong Kong, Portugal, Singapore, Bangladesh, Israel, Greece, Kuwait, Switzerland, Chile, and Austria to mention just a few.

In fact, assuming this data is correct and Bitcoin was classed as a country it would be ranked 39th in the world according to the official list of countries by power consumption provided, interestingly, by the CIA. Yes, *that* CIA. (Note: there are other lists, but they are not dissimilar)

Just the mere possibility of that is astonishing, but the reality is that it could be even higher. Recently, a note appeared on the official Bitcoin Energy Consumption Index that simply read:

Study reveals Bitcoin's electricity consumption is underestimated … (August 2020).

Following the link provided leads to a new report that concludes that, as of the end of 2019, the conservative estimate was actually 87.1 TWh of power, not 42 TWh as shown on the chart. By that reasoning and using the same formula, does it mean consumption is actually closer to 150 TWh today?

If so, that puts the Bitcoin network as the fourth biggest consumer of power on the planet. Bitcoin would therefore consume more power than every other country except the entire European Union combined, the United States and, of course, China.

Could this really be the case, and if so, what does it mean for sustainability and Bitcoin's future?

Whoa, there

Before we get all carried away with this revelation, it should already be clear that there is significant doubt over just how accurate this data is.

After all, if countries themselves are unable to provide good, solid data for their own consumption figures, how can we possibly say with any accuracy what the true consumption of a disparate bunch of machines spread across the planet really is, especially when we don't really know quantities or specifications?

However, even as a proponent of the ideology and practicality of Bitcoin, I will absolutely concede that it is a) "a lot" and b) set to consume more power as time goes on—at least for now, a point we will revisit shortly. We need to examine what happens next, but since we have no universally agreed data as a starting point it is perhaps more useful to look at trends and take a broader view.

In fact, it may be the only way.

There's Power … And There's Power

When breaking down the "a lot" figure, there are a few caveats that we need to mention, specifically different types of energy classed as renewable, stranded, or unsold energy.

While it's usually claimed that "most" Bitcoin mining is carried out via renewable energy (hydro, wind, solar, etc), the University of Cambridge's September 2020 report challenges this—at least partially.

It certainly agrees that some 76% of miners use renewable energy as "part of their energy mix" but states that, overall, only 39% of total energy is genuinely from renewables, most commonly hydroelectric power.

This is to do with the allocation of hash power—while the Asia-Pacific (APAC) region still produces most of it, their percentage of renewable energy is only 29%, whereas it's as high as 70% and 66% in Europe and North America respectively where hash output is lower.

However, since the world is, generally, moving towards renewable energy at an ever-increasing rate, it is clear that this percentage will increase over time, but it's hard to say what timescale is needed to make the difference.

In addition, there are some circumstances where there are some directly positive outcomes of Bitcoin mining.

"Stranded" energy, for example, refers to the energy that would otherwise be wasted or unused, most commonly gas that has to be deliberately flared due to lack of pipeline capacity or is simply located where it is not cost-efficient to transport.

In those cases, such as the mining farms created in Texas and North Dakota for exactly that reason, there is an unquantifiable benefit of avoiding that action.

Finally, there are areas in the world where too much power is produced and there's no efficient way of transporting it anywhere or to reduce the power output in a way that makes economic sense. This is common, for example, around hydroelectric dams.

Here, power is sold very cheaply to mining operations on the condition they locate themselves in these areas. This creates a win-win for the provider, consumer, AND environmentalists since no extra power is technically produced. Mining farms are one of the few industries that can operate pretty much anywhere as long as an internet connection is available.

However, in truth, there is currently no way of quantifying the beneficial aspects against the obvious negative ones of coal or oil-fired power generation, but we can safely assume that the former is much smaller than the latter and the net effect is still a negative one.

The bigger question would have to be this: if ALL Bitcoin mining was done on a renewable energy basis, would that ultimately be considered a satisfactory outcome for all concerned? For some, yes, but for others, no.

Even without the power issue, there are still environmental concerns. ASICs (the machines used by Bitcoin miners) can only do one task and when they reach life, they are not easy to recycle, if at all. Almost certainly, hundreds of thousands of these machines have found their way into landfills with the promise of more to come.

Not only that, but mining rigs produce a lot of heat and either need more power to cool them (this is, in fact, a significant part of the quoted power consumption figures) OR they need to be located where it is cold.

In the latter case, could there be a long term environmental issue caused by dumping a giant heater in a cold place?

Logic would dictate that there certainly could be.

It seems the problems for Bitcoin keep mounting up.

But, of course, we've been here before.

Deja Vu

Any "gold rush" for new tech leaves a path of destruction behind it. It has always been the case and probably always will be.

Whilst trying to think of a solid analogy, I decided (after rejecting many other possibilities) the most appropriate one was probably the PC itself.

After all, the explosive growth in PC use is still within living memory, the global tech development race was similar and these items were all designed to be used as tools to support and use the main non-sovereign, non-physical product that was initially hard for people to get their heads around—i.e. the internet.

According to historical data collated by Gartner and presented on Wikipedia, between 1996 and 2003 some 922 million PC units were sold globally. I chose these dates because this was the era of proper 'first-gen' home computing, where connectivity to the net, such as it was, was possible, but difficult. After 2004, ADSL lines and other solutions started to become widely available and the tech began to change accordingly.

That's a huge amount of boxes, but how many of those do we still use today? The answer, almost certainly, is none. Well, apart from the one my mum keeps in her spare room to play solitaire on occasionally.

Almost a billion units have been thrown away and, since these were the days before recycling, they also probably made it to the landfills too. And what about the billion or so modems we produced in the days before ADSL? Now all gone too. It was expensive both in terms of environment and dollars, but things have settled down considerably since then.

Last year just 261 million units were shipped compared to the peak in 2011 of 353 million, a drop of 102 million units in a trend that is still falling. We use other devices these days instead of just computers and we tend to upgrade far less frequently anyway as the gap between hardware and software capability has narrowed.

The PCs we do still use are far more energy-efficient than their ancestors, especially in terms of monitors, so as the number of connections to the internet has increased over the years, it's likely that power consumption hasn't increased proportionately.

My view is that, over time, we will see the same with Bitcoin.

Right now, we're in an arms race where even a slightly better machine is sought after at almost any price in the same way a 66Mhz processor was favored over a 33Mhz processor in the early, frantic days of computing.

But like PCs, over time this will settle down and many of the issues will be resolved, something I may seem overly confident about.

Here's why:

The Future and Context of Proof of Work

Although I am a self-confessed optimist on, well, pretty much everything really, I am also quite cynical about certain aspects of human nature. The bottom line is that, collectively, we just won't care enough about things like the environment unless a) it affects us directly and b) it makes us money.

The trick, therefore, is to align global needs with selfish needs, or, to put it another way, to ensure that making money happens to coincide with things that are beneficial to the planet.

And although Bitcoin seems to be the opposite at first, from a long-term perspective there's a natural alliance that could actually turn out to be very beneficial for all parties concerned.

Making money as a Bitcoin miner is all about efficiency and, as time goes on for technical reasons, this becomes more and more important.

Right now, we're using the mining equivalent of the huge, extremely inefficient engines that were prevalent in American cars in the '70s and got 8–10 miles to the gallon. After all why not? Fuel was cheap, plentiful and the environment (whatever that was) was fine with it, right?

But you'd never buy a new car like that now, partly due to a change in social awareness, but also because it would be too expensive to run. As soon as it becomes more about money, we pay attention.

Right now, new, sleeker tech is being developed. Each generation will produce more hash power for less energy and as these lines converge—as they always do—it simply won't be cost-effective to run older, hotter, juicier equipment that needs to be frequently upgraded.

As time goes by, and even as Bitcoin's hashrate and influence increases, I bet you any money (*in bitcoin please, not dollars*) that its carbon footprint doesn't increase in proportion to that growth.

This will be partly because of the equipment itself, but also due to the fact that earlier this year a whole swathe of reports, such as this one, came out confirming that solar and wind are now the cheapest forms of energy production on the planet. Renewables, like Bitcoin, are here to stay and will now grow faster than any other sector as a result.

About time too.

A Net Reduction?

I further bet you based on what we've seen historically, applications of Moore's Law (the concept rather than the technical absolute) and the fact that we humans, collectively, are predictable when it comes to money, that this power consumption will plateau and *even fall* eventually.

It's a bold statement and it's definitely not something that will happen next year—although certainly could within the next decade—and I'm happy to go on record with that forecast.

There's always a counter position of course, and the most relevant ones, in this case, is the LED light argument, covered in this article. The crux of this report is that overall power consumption in lighting has increased, even though much less power is now required to produce the same amount of light.

I'm sure that's true as population and applications have grown on a global level, but when I switched all my lights at home to the latest-gen LED lighting, I noticed an immediate and significant reduction in both power consumption and energy bills as well as a far superior light. I have also never had to replace any bulbs as yet, something I definitely would have done several times over by now using the old, hot, inefficient tech.

I think we can safely assume Bitcoin miners, on an individual level, will consistently seek to do the same. After all, their livelihoods depend on it.

And there's also one more consideration worth noting.

It's easy to overlook that the traditional banking and global payment processing sectors also require enormous amounts of power to run, although it's even harder to calculate what these numbers might actually be. There's no single network to base assumptions on and there's literally thousands of independent companies and organizations that are often not exactly forthcoming about their carbon footprints.

However, we can safely assume that this power consumption, globally, must be *at least* what Bitcoin's requirement is.

Bitcoin, in my view, will never replace that existing banking system, nor should it, but there is at least the *possibility* of reducing it in the long term. For example, every transaction that takes place over the Bitcoin network is, mathematically speaking, a transaction that does not take place on the traditional network.

It's a very simplified view, but on a big enough and prolonged enough scale, this could make a difference, especially when it comes to decisions about power-hungry physical premises and data centers by incumbent financial institutions.

The size of that impact may ultimately be tiny or it may be significant, and my guess is that it will be somewhere between the two. On a global scale, this is therefore a redirect of power resource rather than a requirement for more, and that can be considered a benefit of the Bitcoin network.

The Bottom Line

I think Alex De Vries's work is actually very good as he embarked on a difficult task with little hard information. It has limitations, but these are clearly explained through caveats.

However, I disagree entirely with his prognosis as it assumes that what is happening right now will be the same going forward. I propose that this will not be the case because the parameters will change, the tech will change and as for the rule book—well, we haven't even got round to writing that yet.

Although I have no doubt that De Vries's work is right in the short term while we power through this crazy growth-spurt-cum-gold-rush phase of Bitcoin's lifecycle, the adoption and management phases will see a different, more measured approach where power efficiency will become ever more important.

This means there will be an even more absolute and direct link between the power consumption of the miners and their standard of living. You can bet, therefore, they will continually seek out the lowest cost energy sources (already proven to be renewables), and the most efficient, lowest heat-producing machines that have the best longevity. Y'know, a bit like my LED lights.

They will also do this with a level of motivation that would outweigh anything in the traditional banking sector because they would be directly incentivized to do so.

And when individual needs align with things that are good for the planet, you know they will get done, even if it's not for the reasons we'd prefer.

Still, we'll take it.

And so will Mother Nature.

Can Blockchain help Stop Human Trafficking

23

Written by Jillian Godsil

Human trafficking is a multibillion-dollar business and it is growing exponentially. Consider the fact that a kilo of drugs can only be sold once only but a human being can be trafficked over and over again. It's all in the maths. 40 million people are trafficked each year with half being women and children, and most sold into the sex trade. This is modern slavery.

While researching this story I spoke with a number of people active in highlighting the issue or seeking to put a stop to this heinous crime. I had imagined that human trafficking belonged solely to developing countries where human life was considered cheap – not a comment on my part, I may add, but an observation. To my shock, I discovered that human trafficking happens under our noses in well-developed countries and includes people of all nationalities including the host nation. In fact, if a child runs away and is on the streets in any city, any western city, experts say that the odds are they will be trafficked if not found in 48 hours. 48 hours? That is in just two days.

It's also a scientific fact that we humans are never more than six feet away from a rat, our faithful shadow. The same can be said for trafficking human beings – they are all around us, but we have not been educated to open our eyes.

A charity in Ireland called MECPATHS (Mercy Efforts for Child Protection Against Trafficking with the Hospitality Sector) was founded in 2013 as a non-profit organization to work directly with the hospitality industry to prevent child trafficking. One in four victims of human trafficking worldwide is a child and in COVID this is even worse as children in danger can be locked down with their abusers or spend more time online and risk contact with traffickers.

The definition of child trafficking is the 'recruitment, transportation, transfer, harboring or receipt' of a child for the purpose of exploitation under the United Nation's Palermo Protocol, and a child is defined as any person under the age of 18. The United Nations regards all trafficking as modern slavery.

Trafficking exists undetected for two main reasons. The first is a lack of education on what a trafficked person might look like, how they might behave, where they might be found. The second is our misplaced emphasis on privacy.

Patrick Baker, an independent Irish radio documentary maker, spent a year working with MECPATHS to record their work and produced an impactful documentary called Do Disturb. In it, she explores what we need to look for and how MECPATHS works with the hospitality sector to make them aware.

"I fell into this story while researching another, and was shocked to discover not only is human trafficking present in every village in Ireland, but we here in Ireland have been downgraded in the annual Trafficking in Persons Report published by the US Department of State to Tier 2-watch list. The need for public awareness is paramount; the statistics are frightening and overwhelming."

Detecting child trafficking is not a clear science, but a series of observations combined can help illuminate such slavery. MECATHS educates people working in the hospitality sector. Porters are taught to check if young adults or children arrive without luggage. Receptionists are taught to see if the child looks upset, if the body language looks off, even if they do not resemble their 'parent'. Do they have identification? A passport? Then maids are taught to note if extra towels are ordered for a room. Night porters are taught to question multiple visitors to a room. It is not just one thing; it is a series of things and what is the worst outcome that can happen if an incorrect call is made? Egg on the face? Perhaps, but surely that is so much better than having a child trafficked under your nose.

In the radio documentary, a brave young Irish woman explained that her father trafficked her for sex from age four until eight. She argued that someone might have seen something wrong, spotted something amiss. She called out for people to 'twitch that curtain' and be nosy. And this is the second point that lets off the purveyors of human trafficking.

We are taught not to interfere in other people's business, in their marriage, in their home. It is only recently in Ireland that domestic abuse is being called out, people are calling it out, it is no longer good enough to say what happens behind closed doors is their own business. It feels alien to tell tales but when lives are at stake then we are encouraged to be that whistleblower.

There is a third reason why trafficking is growing – the lack of prosecutions since 2015. Globally, these have fallen by 42% and in Europe, prosecutions have fallen by 52% leaving the perpetrators of these crimes to conduct their business with impunity.

If awareness and education can help stop trafficking, how can technology help?

One way can be to limit access to the web. In February 2020, the Irish web hosting company Blacknight joined industry giants such as GoDaddy, Nominet, and Amazon to agree on a common framework to address abuse of the DNS by criminals. This voluntary agreement identifies four types of web content abuse which include child sexual abuse materials and human trafficking. If witnessed, the 48 companies so far signed up have agreed to disrupt this content without waiting for a court order.

But how can trafficking be prevented at the source? This is an issue that led entrepreneur Jax Harrison to set up two companies: a for-profit to develop and combine techniques to help children at risk called Innovation For Good and a not-for-profit to implement the technology and education called StopChildTraffic.org.

Originally from New Zealand, Jax worked in Peru for many years and witnessed first-hand the fate of street children in a time of terrorism and political unrest. She was involved in helping to establish an NGO for a local orphanage where parents would surrender their children to keep them safe from being trafficked, into sex, or as child soldiers.

This sparked her lifelong interest in working to save such children but to her horror when she traveled to the US and settled in Kanas City, Missouri she discovered child trafficking was very much present there too, in fact, she discovered that Kansas City is considered an origination city – from which children were trafficked across the country.

Now she uses technology, and notably blockchain, to develop technology to protect children. Her starting point is prevention and she has worked closely with the UK government to provide safe protocols for children accessing online services.

"That also includes educating parents what to look for in their child's behavior and if your child is playing an online game, chances are they have already been approached in the accompanying chat rooms. There are some basics – parents are meant to parent and not to be their child's friend. Simple actions like once a week looking at their phones, checking where they have been, and having access to their passwords."

That is just one element, Jax is bringing together CEOs of technology companies to use AI and blockchain to map the locations of vulnerable children and potential traffickers.

"We follow a similar trajectory used to track terrorists. It's all about disrupting the supply chains, with the slim numbers of people being prosecuted we need to break the chains so they can't pass on the child."

Another key element is identity. A child or person who is trafficked is most likely denied their identity or may not have any. Likewise, they do not have access to money and are largely unbanked. She is working to develop a platform that allows an undocumented person to claim their identity and attach it to a self-sovereign wallet.

"Identity and financial access are key to keeping people safe, recovering trafficked people, and helping them get their lives back."

For example; Innovation for Good and StopChildTraffic are working with WyOfi and Secours.io to help provide access to good digital identity and to finance.

WyOfi are the Innovators bank and as a Wyoming state-chartered Special Purpose Depository Institution (SPDI), WyoFi is a bank with purpose – enabling innovative companies to operate through their compliant financial services on a robust fintech platform in a thriving global ecosystem. One of the ways they do this is through their proprietary blockchain protocols which provide frictionless transfer of all types of currencies i.e from Krypto to Dollars. They do this in a seamless user experience enabling one currency to transfer to a preferred currency providing a frictionless transfer from sender to receiver. This opens up opportunities for at-risk populations to use all forms of digital currencies and transfer options.

Secours.io is the first global, cryptographic public safety network based on the TrustOverIP Foundation and Sovrin Foundation governance architecture.

A new hybrid platform whose most powerful imperative is trusted digital relationships, verified credentials, secure data transactions, and AI. This unified global safety infrastructure empowers a new digital mechanism of trust between citizens and public safety.

This risk mitigation, loss control platform which can take the form of a digital wallet allows integration of any application, phone, or

IOT device, freeing others from having to create this pivotal public safety infrastructure. They bill themselves as the " Intel Inside" of public safety.

"It is imperative that we use all forms of cutting-edge technology for the good of humanity and to prevent harm to those who are vulnerable to these heinous crimes. It is only by working together that we can break the supply chains and save millions of children from being trafficked," *says Harrison.*

Jax recently served on the UNITED NATIONS -SF Board as the vice-chair of the Women's Committee, has advised the UK Government on keeping children safe – specifically in the area of ONLINE HARM, is Chair of Communications for AAUW Marin, (a national women's rights organization), sits and advises on the San Francisco Human Trafficking Coalition Board, The San Francisco Mayors Council for Human Trafficking, and the San Francisco Youth Human Trafficking Council is a stakeholder for the NZ Digital Council Alliance, and a member of the NZ Women in Identity Group, ArchAngels Investors and of the Trust over IP Foundation.

Crypto in The Great Reset

24

Written by Mark Bailey

The World Economic Forum (WEF) has started pushing something called 'The Great Reset'. The idea is that society must transform in response to economic turmoil triggered by the COVID-19 pandemic, in anticipation of further economic turmoil related to climate change. More details of The Great Reset will likely become available after the WEF meets in Davos in January 2021. But the information that is already available suggests that The Great Reset is a very big deal. The Founder and Executive Chairman of the WEF, Klaus Schwab, wrote:

"COVID-19 lockdowns may be gradually easing, but anxiety about the world's social and economic prospects is only intensifying. There is good reason to worry: a sharp economic downturn has already begun, and we could be facing the worst depression since the 1930s. But, while

this outcome is likely, it is not unavoidable. To achieve a better outcome, the world must act jointly and swiftly to revamp all aspects of our societies and economies, from education to social contracts and working conditions. Every country, from the United States to China, must participate, and every industry, from oil and gas to tech, must be transformed. In short, we need a "Great Reset" of capitalism."

So The Great Reset is a reset of capitalism, wherein every industry is transformed. This sweeping change involves three main priorities:

- Steer the market toward fairer outcomes.
- Ensure that investments advance shared goals, such as equality and sustainability.
- Harness the innovations of the Fourth Industrial Revolution to support the public good, especially by addressing health and social challenges.

On the surface, this all sounds great. Considering the WEF's embrace of blockchain and other emerging technologies, it sounds almost possible. But after reading more about The Great Reset on the WEF website and perusing *Time Magazine's* surprisingly extensive coverage of the subject, I noticed some troubling patterns in the ideas being presented. And when I went looking for alternative viewpoints to mainstream presentations of these ideas, I found few items of substance.

The Hill, a notoriously conservative publication, ran an opinion piece that treated The Great Reset as a grave socialist threat. A blogger named Tessa Lena wrote a long post titled, 'The Great Reset for Dummies" which was somewhat more interesting. This piece considered the implications of creating a digitized global inventory of all things, the disappearance of privacy, and the possibility of a new global feudalism arising as a result of The Great Reset. Also, for reasons I couldn't ascertain, it treated crypto in general as a major threat to freedom. Lena's evidence seemed to boil down to this Microsoft patent for turning human activity into cryptocurrency, which I find entirely nonthreatening.

A more thoughtful look at the role of crypto in The Great Reset was published by Raoul Milhado in the *Coin Telegraph*. Here's a quote from that article:

"The Great Reset will be a revolutionary upheaval. It will create many hardships but also opportunities for companies, which recreate financial products in a decentralized fashion on blockchain, and for individuals, who no longer must rely on the traditional financial system. ... The Great Reset might come down to central bank digital currencies versus the cryptocurrencies of the people — centralization versus decentralization."

Will The Great Reset come down to centralization versus decentralization? I doubt it. But I do think The Great Reset might attract significant corporate backing. If it does, that might mean disruptions in some major industries.

Healthcare could be impacted, perhaps by the introduction of new medical surveillance standards. Carbon trading markets could also get shaken up. Fossil fuel divestment seems likely to accelerate under Great Reset policies. And The Great Reset's emphasis on fairness might translate into subsidies or social welfare programs that do more for bloated bureaucracies than for people in need.

There are obvious advantages to employing crypto and blockchain technologies to address social welfare challenges. But although the specifics of The Great Reset are not yet public, recent history suggests that its international efforts might shy away from public blockchains. When the United Nations World Food Programme began using blockchain to distribute aid to refugees in Syria and Jordan, it used a variant of the Ethereum blockchain in combination with a traditional United Nations database to do so. This 'variant' consisted of a private fork of the Ethereum codebase customized by the engineering firm Parity. Instead of being given private keys to their accounts, aid recipients accessed their accounts with eye scans done by a proprietary device.

To be honest, I don't think even the WEF has the power to reset capitalism. And I am skeptical whenever an organization like the WEF starts talking about issues like fairness and sustainability. If the WEF is serious about creating a more equitable world, perhaps it will embrace crypto in ways never before seen by an institutional actor. But my guess is that the WEF will embrace blockchain in limited areas like supply chain management, and do everything it can to avoid giving people all the messy freedom that comes along with cryptocurrencies.

Physical Security as Crypto Goes Mainstream

25

Written by Mark Bailey

Armed robberies have become common in my Minneapolis neighborhood. Every day, people within a mile of where I live have their cars or wallets stolen at gunpoint. The robbers always take the victim's phone. And when the victim has Venmo, it has become common for the robbers to use Venmo to transfer themselves hundreds of dollars or more.

New Risks as Companies Adopt Crypto

Along with PayPay, Venmo appears poised to integrate cryptocurrency payments into its platform in 2021. While few criminals are yet sophisticated enough to use specialized crypto wallets to pilfer funds from a phone, the criminals where I live can definitely use Venmo. If Venmo is connected to a crypto wallet, the tokens in this wallet would be an easy target for these criminals. I fully expect to read about a crypto robbery on my neighborhood watch app in the next year.

Personally, I don't use Venmo and only rarely use my phone for crypto business. I also tend to leave my phone at home when I'm out and about doing everyday things, so it's unlikely the device would be lost if I were mugged. But I'm generally risk-averse. Not everybody is.

Since last spring, criminals have become very active in my city. There've been over 4,600 violent crimes in Minneapolis so far in 2020. For a while during the George Floyd protests, the city was completely lawless. In the wake of those protests, political wrangling has limited the number of actively patrolling police officers, which has not encouraged a return to law and order. I expect at least another year of high crime where I live.

At some point, as crypto goes mainstream, street criminals are going to start chasing it. There have already been reports of armed assailants going after cryptocurrency in New York, the UK, and the Netherlands. But we are not powerless against this trend. Measures can be taken to reduce the risk of loss of funds in the event of a robbery.

Basic Physical Security Measures

This *New York Times* article describes some of these measures. Creating a decoy 'duress wallet' with limited funds in it gives you something manageable to hand over to would-be robbers. Using multi-sig wallets prevents funds from moving with just one party's approval, which makes funds hard to steal. Refraining from talking about cryptocurrency holdings in public limits possible exposure to criminal attention.

It isn't yet clear how crypto will be integrated with mainstream mobile payment systems. Venmo doesn't seem likely to accommodate multi-sig wallets. But limiting the funds in any Venmo-attached wallet would effectively make it into a 'duress wallet' that could be given up in the event of a robbery. And just as there are ways to limit the attack surface of a website, there are common-sense practices that make a physical robbery less likely.

These practices may vary depending on the habits of the criminals in your area. In Minneapolis, where over 3,000 cars have been stolen this year, one of the most important measures that can be taken is to keep car doors locked at all times. And if you see a car or person blocking the street, don't stop to see what's going on. Back up and get out of there!

As a pedestrian, situational awareness is even more critical. In my neighborhood in broad daylight, robbers cruise the streets in stolen cars, looking for distracted victims. So don't appear distracted. Avoid being on your phone or wearing headphones. If possible, avoid carrying a phone or purse in your hands. And if you see someone exit a car with no license plates to approach you, leave the area as quickly as possible.

Another thing criminals in my city have started doing is going to bars, coming up with an excuse to hold someone's phone, and then using that phone to transfer themselves money using Venmo. The solution to this problem is to never let a stranger hold your phone. Not even for a moment to show you something on Instagram. That thing on Instagram is probably not worth the risk of loss of funds.

Security culture is baked into the cryptocurrency community. As this community grows rapidly, fueled by mainstream adoption of crypto by companies like Paypal and Venmo, new security vulnerabilities will emerge. Extending security culture from the virtual to encompass the exposure of the physical limits to these vulnerabilities. Accounting for street crime in our security thinking can help keep us safe and happy.

You Don't Need More U.S. Dollar to Spend, You Need More Bitcoin to Save

26

Written by Sylvain Saurel

> *"You can always make more money, but you can never make more time."*

The source of this quotation is still very mysterious. I have heard it throughout my life, but I have never been able to find out who it was originally from. Anyway, I find it full of meaning.

This quote makes even more sense when I see some people going on a frantic quest to earn more and more fiat money. While we are experiencing great monetary inflation, some people still believe that the U.S. dollar is all-powerful.

All these people think about is earning more and more U.S. dollars to accumulate more and more.

Accumulating More and More U.S. dollars Will Not Make You Happy

I still wonder how these people don't realize that the U.S. dollar they cherish so much will never bring them true happiness in life.

The first thing to understand is that the U.S. dollar was not designed to be saved. The U.S. dollar was designed to be spent. It is the model of the system at the center of which is the U.S. dollar. To force you to spend all your hard-earned money, the central bankers keep increasing the money supply.

This endless inflation of the U.S. dollar, and other fiat currencies, induce an equally endless devaluation of your purchasing power.

In 2020, the purchasing power of $1,000 in 1971 is no longer even $150.

This drop of more than 85% in 49 years is quite phenomenal when you think about it.

The most unbelievable thing is to think that to keep the same purchasing power at $1,000 in 1971, you would have had to transform this amount into $6,426 in 2020. This could only have been done by making investments in different areas, but certainly not by saving your money in the bank.

Unfortunately, a majority of people do not have the opportunity to buy real estate, invest in the stock market, or buy gold to protect themselves from this monetary inflation.

The U.S. Dollar is at the Center of an Unfair System

The current system is unfair because it produces a slow but inevitable impoverishment for a majority of people.

While the richest 1% of the population continues to see their wealth appreciate, the poorest 50% live in increasingly difficult conditions.

Between 1989 and 2018, the richest 1% of the United States saw their wealth increase by more than $21 trillion. At the same time, the poorest 50% saw their meager wealth decrease by nearly $1 trillion.

The U.S. dollar is a currency that does not give equal opportunities to all U.S. citizens.

It is the same with all fiat currencies in the different countries of the world. However, hundreds of millions of people work harder every day to try to earn more U.S. dollars.

Since they can't save their fiat money, they are doing the only thing the system expects of them: they are spending more and more money on totally useless things.

The U.S. Dollar Locks You Into a Vicious Circle of Impoverishment

The consumer society is very strong to persuade you that you have to buy those things that are useless in the end. Since you no longer see any point in saving, you end up giving in.

This leads you into a vicious circle that will prevent you from achieving happiness. I am talking about true happiness here: having enough time to do what you want in life.

The U.S. dollar doesn't give you access to this most precious resource in the world: time to live your life as you wish.

So clearly the solution is not to earn more U.S. dollars to spend. You have to look for a way out of this vicious circle that impoverishes you irretrievably over time.

Bitcoin is Your Way out of this Vicious Circle

Bitcoin is the exit you need. Officially launched on January 3, 2009, Bitcoin is a savings technology that gives you control over what you own. Once you own the private keys associated with your Bitcoins, you are the only master on board.

No one can confiscate your Bitcoins, or prevent you from using them as you wish.

Therefore, Bitcoin gives you the freedom to live on your own terms. You can buy one Bitcoin in 2020 with the guarantee that it will still be one Bitcoin on 21 million in 10, 20, or 50 years. This guarantee is given to you by the fact that the Bitcoin offering is hard-capped at 21 million units.

Bitcoin has been steadily increasing in value since its launch. From $0.125 in October 2010, Bitcoin's price has grown to over $13K in October 2020. No asset in the world has done better over the period. If you had bought $1,000 worth of Bitcoin in October 2010, you would have had the equivalent of $108 million in your possession when October 2020 came to an end.

As of November 20, 2020, the situation would be even better since the Bitcoin price is in the midst of a bullish rally and has exceeded $18K.

The more U.S. dollars you earn, the more Bitcoin you will be able to buy if you fully believe in its monetary revolution.

This will help you be happier. It is not the possession of Bitcoin itself that will make you happier, it is what it will allow you to do. Indeed, you will have more time at your disposal. Time is the most precious thing on Earth.

Conclusion

Bitcoin is the scarcest human invention in history. More Bitcoin will help you get more time to live your life the way you want. Bitcoin will help you achieve true happiness in your life.

Few people realize it yet, but Bitcoin was designed to give every citizen the power to take control of their lives. When you buy Bitcoin, you decide to become your own master. You can then maximize your chances of achieving the happiness we all aspire to in life.

Bitcoin Won the U.S. Presidential Election

27

Written by Mark Helfman

You may think the U.S. presidential election is under dispute.

It's not.

The president's team claims fraud and tweets all sorts of evidence, but they don't present any of that evidence in court. Even their "independent advisors" found no fraud. I suspect President Trump knows the fraud claims won't stick, and in any event, he's told his administration to let his successor, Joe Biden, take office.

Unless something changes drastically, the U.S. will have a new president in January.

If you believe people who worked for Trump, the president ordered his administration to "go after" bitcoin.

His Treasury Secretary called crypto a national security issue. His law enforcement agencies worked to de-privatize transactions. One of his former financial regulators bragged about conspiring to kill the 2017 bull market.

You can expect the Biden administration to take a more favorable view. Not sure it could get any worse.

Bitcoin ETF, anybody? Safe harbor for small investors? Sane tax guidance and regulations? U.S. federal funding for blockchain projects and blockchain-for-government efforts?

Those things are a lot more likely with this administration than the outgoing one.

Some of President-elect Biden's biggest donors have strong ties to crypto. Silicon Valley has a lot of influence in Democratic Party circles and Wall Street has its hands in both parties' pockets.

If you believe the rumors about who Biden will appoint as cabinet officials and top advisors, you will find a slew of pro-crypto, pro-blockchain appointees, including his likely securities regulator and Secretary of Labor. In a boon for all those who love BRRR memes, His pick for Treasury Secretary is an inflation dove who has historically supported low-interest rates and cheap money (her stance on crypto is not as clear).

Also, the Democratic Party skews younger, both in absolute numbers and as a proportion of people who register as Democrats compared to Republicans. Many surveys show young people have more positive opinions of cryptocurrency than old people.

Still, presidents can only do so much. They don't make laws. That's Congress's job.

U.S. regulations come from financial laws that are 50 to 86 years old. Until those laws change, the president has limited power. He can issue regulatory approvals and guidance, perhaps sign a few executive orders. Significant actions but hardly game-changers.

So it really depends on how much emphasis the new administration puts on cryptocurrency and pushing Congress for crypto legislation.

A New Crypto Coalition?

If we're speculating on hypotheticals, I can imagine an interesting dynamic could play out within the Democratic Party.

This party has a lot of pro-Wall Street people and a lot of anti-Wall Street people. The pro-Wall Street people want institutional and government involvement to clean up the cryptocurrency industry and make money from it, while the anti-Wall Street people think cryptocurrency can destroy what they see as an unfair system that rewards rich people for ripping off poor people and small businesses.

Maybe they'll find kindred spirits with Republicans that value free markets, individual liberties, small government, and low taxes. As a result, the U.S. will finally get sensible cryptocurrency laws.

Or not. It's fun to speculate.

In any event, cryptocurrency is a bipartisan issue—namely, neither party cares about it much. It may rise in prominence as the market attracts more investment and prices go up, but it's anybody's guess how Washington will respond.

But Won't Biden Lock Down the U.S.?

We'll see.

Until January 20, 2021, Biden can't do anything. A lot can change in two months.

Whatever happens, I doubt it will affect crypto. The entire market doubled during our previous lockdown and big money gobbled up tons of bitcoin. It continues to rise despite lockdowns in some parts of the world and new COVID-19 restrictions in some parts of the U.S.

I can't see why another lockdown would get a different result.

Let's hope we don't have to find out.

In Conclusion, None of This Matters

While it's nice to think the Biden administration will bring sanity to the U.S. regulatory and legal treatment of bitcoin and the cryptocurrency industry, it doesn't matter for prices.

Bitcoin is a global commodity and Wall Street has built the legal, operational, and regulatory infrastructure to move lots of money into the markets. Soon, greed will take over.

It probably already has.

Bitcoin's price will rise no matter what the Biden administration does, as will the entire crypto market.

But prices will not go up forever.

Once we get the inevitable crash that comes at the end of each market cycle, we will need good laws, fair regulations, and reasonable protections for businesses and investors. Only then can these open, permissionless financial systems grow, thrive, and deliver the freedom, innovation, and efficiencies everybody expects they will.

No industry thrives in chaos. Speculation alone can't sustain cryptocurrency.

Let's hope the Biden administration cares about that, too.

MicroStrategy is to Bitcoin as Roger Bannister is to the 4-Minute Mile

28

Written by Jason Deane

Once the Barrier is Broken, the Flood Gates Open

Recently, and without warning, a piece of news hit the financial reporting outlets without fanfare or fuss and caused a collective gasp among the cryptocurrency community.

A Nasdaq-listed company, Microstrategy (MSTR) had nonchalantly announced that it had purchased 21,454 bitcoin for $250 million, including all fees and expenses.

From an outsider's perspective, it did it as casually and calmly as someone getting in from work and announcing that they'd got some gas on the way home. It could easily have been preceded by the words "Oh, by the way ..."

I'd heard of Microstrategy through the reputation of course, but other than the fact I'd seen it on the Nasdaq and its logo looked a little like Microsoft did in the nineties, I knew little about the company itself.

It's no coincidence, of course, that the first company to secure significant amounts of bitcoin for its "treasury reserve strategy" is a company entirely dedicated to forward-thinking research and business intelligence. In actual fact, it makes perfect sense that it would be.

These are very sharp people who have access to some of the best analytical tools in the industry. In fact, they *created* a lot of them, and many of the Fortune Global 500 now use those very same tools for their own applications.

So, if a future-focused company whose work is based on modeling likely outcomes for the future using cutting-edge data suddenly buys a lot of bitcoin, what does that tell us about what's coming next?

Cool as a Cucumber

Micheal J Saylor, CEO of MicroStrategy, used deliberate language in his company's press release that made the announcement on Aug. 11, 2020.

You can't help feeling he's the cool guy who's walked into a crisis somewhere, slowly removed his shades, and causally announced the rescue plan in such a way that everyone immediately listens and follows.

This statement, for example, is what many Bitcoiners have been purporting for some time:

"This investment reflects our belief that Bitcoin, as the world's most widely-adopted cryptocurrency, is a dependable store of value and an attractive investment asset with more long-term appreciation potential than holding cash"

He goes on to explain that the current financial landscape is "creating long-term risk for our corporate treasury program" and singles out, in particular, quantitative easing and political uncertainty as key reasons for that thinking.

None of these statements, in and of themselves, are a surprise for economists and analysts like myself, who work on or follow this stuff on a daily basis. What is more interesting is that out of all the asset classes available for the company to purchase, they chose bitcoin.

The press release implies that consideration was given to alternatives, but they seem to have been dismissed outright early on. Saylor, still as calm and collected as ever, simply states the following:

"We find the global acceptance, brand recognition, ecosystem vitality, network dominance, architectural resilience, technical utility, and community ethos of Bitcoin to be persuasive evidence of its superiority as an asset class for those seeking a long-term store of value. Bitcoin is digital gold-harder, stronger, faster, and smarter than any money that has preceded it. We expect its value to accrete with advances in

technology, expanding adoption, and the network effect that has fueled the rise of so many category killers in the modern era."

It's nothing short of bold, delivered in a matter-of-fact fashion reminiscent of a judge handing down a verdict for a long and tricky case. This is a man who is clearly sold on his convictions.

And then there's the investment level itself. This wasn't a spare few thousand earmarked for a new, fancy office printer that they decide they could do without for a while, this was an investment of a quarter of a billion dollars of funds in a publicly-traded company. Funds for which they know they will have to answer for.

This was planned, processed, and delivered in meticulous detail. This was deliberate and unapologetic.

So, What's Next?

Like anything that seems impossible, it only is until someone does it.

When, for example, Roger Bannister publicly and spectacularly broke the belief that a human being could not run a mile in less than four minutes in 1954, it was not the first and only time it has happened since. Now it is common, almost routine, and there is certainly no fanfare when it is achieved.

In my view, the same scenario is likely to play out with this situation. It's always been possible for any company to buy bitcoin or, for that matter any other asset, and add it to their balance sheet, but the question is, who would go first? Who would risk putting their head above the parapet? What if they were *wrong*?

about *making the needs of the people, the needs of the government.* This is the meaning of representation in government.

When the American colonies suffered great hardships during winters, European powers harbored little concern. Those moments would be felt later across the Atlantic in various forms. Maybe diminished resources or falling incomes spurred action in Europe. The result? Americans could be said to have paid a dual-suffrage in that of the frigid onslaught of the North East (*Mother Nature*) and a second from *Mother* Europe.

Parallels in Modern Governance. Alas, our prosperity as Americans imposes a similar weight upon us. Technology is the new divide and soulless servers the new throne of governance. Centralized data systems immensely accelerate emerging economic and resource divides.

The Power of Representative Voting. Blockchain gives individuals the power to bridge the divides that exist across lifestyles. When voting systems leverage smart contracts, *all the parts of a civilization can coordinate their resources and needs.* For those less familiar with the capabilities of smart contracts, decentralized finance might be a good place to start. Blockchain applications have already been deemed essential for things like logistics, standardization, and networking. Mobilizing resources alone is enough to press for blockchain. Couple that with matching resources with needs, and suddenly, voting power becomes something near miraculous.

Concluding Thought

We need to vote because we know its power. This is also how many feel about blockchain technology. Blockchain without voting is like an unauthorized market. Individuals and small organizations end up either building tools for the greater market or operating in redundancy. Voting without blockchain is failing to take a step forward. Voting on blockchain ascends decision-making processes on par with the transition from daily rush hour traffic to owning your own general aviation aircraft. Yes, we are ready.

Tim May, Cypherpunk

31

Written by Mark Bailey

While working at Intel, Tim May discovered that the ceramics being used in some computer chips were slightly radioactive, which caused data processing problems. After discovering this, he was awarded his own lab and stock options significant enough to allow him to retire at age of 34. May wrote *the Crypto Anarchist Manifesto* in 1988.

Today, this 500-word essay appears to have been somewhat prescient.

In my opinion, here's the manifesto's most important paragraph:

"Just as the technology of printing altered and reduced the power of medieval guilds and the social power structure, so too will cryptologic methods fundamentally alter the nature of corporations and of government interference in economic transactions. Combined with

emerging information markets, crypto anarchy will create a liquid market for any and all material which can be put into words and pictures. And just as a seemingly minor invention like barbed wire made possible the fencing-off of vast ranches and farms, thus altering forever the concepts of land and property rights in the frontier West, so too will the seemingly minor discovery out of an arcane branch of mathematics come to be the wire clippers which dismantle the barbed wire around intellectual property."

May, who passed away in 2018, understood crypto's underlying technology as well as anyone and brought people together to work out the possible implications of this tech in a variety of ways. He even ran a Cypherpunk mailing list in the 1990s that counted Julian Assange as a contributing member. The surveillance state that grew out of 9/11 gave many in this Cypherpunk community 'cold feet', which spelled the end of the mailing list. But many of May's ideas remain as relevant today as they were 20 years ago.

Before his death, *Coindesk* asked May to send them his thoughts on Bitcoin. May replied with a 30-page missive on a variety of subjects related to crypto. *Coindesk* turned this missive into a faux interview for publication. This faux interview was titled, "Enough with the ICO-Me-So-Horny-Get-Rich-Quick-Lambo Crypto."

Privacy vs Dossier Society

Tim May advocated privacy secured by technology. He adamantly opposed the rise of 'dossier society', wherein business and government use computerized files kept on individuals for the purpose of social and political control. In crypto, May argued against all attempts at regulation and KYC policies. *Coindesk's* faux interview suggests he was not at all happy with the state of the cryptocurrency community in 2018.

The whole world is arguably turning into the dossier society that Tim May feared. The coronavirus pandemic has fueled an acceleration of the disappearance of privacy. Companies like Clearview AI now provide corporate and governmental clients with the means to eliminate anonymity entirely. In the US, voices like Rep. Alexandria Ocasio-Cortez apparently favor creating a database of their political enemies for the purposes of blacklisting them from polite society. And China's social credit system appears emblematic of everything May opposed.

Comparing Bitcoin to speech, May considered the disappearance of privacy in crypto transactions to be akin to censorship. "I'm just saying we could end up with a regulation of money and transfers that is much the same as regulating speech," May told *Coindesk*. "Is this a reach? If Alice can be forbidden from saying 'I will gladly pay you a dollar next week for a cheeseburger today,' is this not a speech restriction?"

Recently, exchanges like ShapeShift have started de-listing privacy coins like Dash and Zcash. But even in the march towards dossier society, Bitcoin transactions remain irreversible, which May staunchly supported. And, throughout the cryptocurrency community, developers continue to advance peer-to-peer projects designed to eliminate unnecessary third-party intermediaries, which May also supported.

Controversy and Legacy

Tim May sometimes expressed his views in deliberately controversial ways. According to his *New York Times* obituary, "In 2003 he wrote that he would welcome a nuclear strike on Washington because it would kill 'a million criminal politicians and two million inner-city welfare mutants.'" Language like that put May at odds with social norms. And promoting the use of cryptography to safely leak government secrets put him at odds with governments.

May's views, however controversial, don't detract from his important contributions to the crypto community. He was guiding serious conversations about crypto when this community was still just an idea being thrown around the Cypherpunk mailing list. When Bitcoin arrived on the scene, it embodied many of May's ideals. And if decentralized internet projects ever fully materialize, they may have Tim May to thank for laying some of the philosophical groundwork on which their tech is based.

Is Bitcoin Just an Experiment?

32

Written by Jason Deane

Anyone who is full time in the cryptocurrency space-like myself-will tell you that most objections to cryptocurrency in general (and Bitcoin in particular) are based around a few key issues.

Generally speaking, these are to do with Bitcoin not being physical and therefore not real (covered in this article) Bitcoin not being backed by anything (covered here) and Bitcoin being no more than an experiment.

Interestingly, most-or even all-of these objections are experienced by the majority of current users, holders, or advocates of Bitcoin at some point on their journey to fully understanding it, and it was certainly true in my own case.

People either take the time to understand the answer to each of these logical questions and ultimately end up adopting Bitcoin, or they don't and simply move on with their lives, staying entirely within the traditional financial system.

Others-such as those with motives not to encourage widespread Bitcoin adoption-will use those very same objections to create doubt in people's minds of the viability of Bitcoin on all levels.

These people usually have a vested interest in the status quo which is, partially at least, threatened by the successful implementation of a non-sovereign monetary system, or have vested interest in a more traditional store of value, such as gold.

However, in all cases, these objections are an illusion and what is most striking about the matter on a macro scale is that money itself-in all forms-is the biggest one of all.

And that includes the cash in your pocket.

In Fiat We Trust ...?

If we go into a shop and buy an item with a note of our local currency, the shop keeper will accept it without any hesitation as long as it is not counterfeit and in a condition that indicates it has retained its value, i.e. it is still recognizable as the unit of exchange it was meant to be.

Most of us don't think about the psychology that is going on behind this transaction.

Fiat currency, that is the notes and coins in your pocket, has absolutely no intrinsic value whatsoever and it is not backed by gold. This is a fact that I believe many of us are probably aware of to some degree or another, but if we ARE aware of it, why do we not question where that value comes from?

The reality is that this perceived value (which is really what it is) comes from our governments simply saying it has value through the Latin decree of "fiat" or "let it be done."

If you want to be more cynical, you could say it has value simply because the "men with guns" say it does. We won't be arguing with them anytime soon.

I find this fascinating because we live in a time when global distrust of governments has never been higher, according to data gathered by the OECD.

There are the usual expectations of course. The governments of New Zealand, Canada, The Netherlands, Norway, Luxemburg, and Sweden have routinely enjoyed a 60%+ trust level for some time, but many countries, including the UK and especially the US, has seen a significant decline in the last few years. In the former case, Brexit is thought to be a significant factor and in the latter, it's usually attributed to the divisive policies of the last administration.

The US, in fact, is now dealing with some of the lowest trust figures it has ever faced, yet very few citizens question the validity of the dollar bills in their pockets. The same dollar bills, in fact, that are issued by that very same government, who, incidentally, also directly control the value of those bills through monetary policy.

It's a paradoxical statement that doesn't stand up to scrutiny. Even hardcore anti-establishment commentators will still trust the very organization they are opposing with their personal wealth by implication of continuing to use it. Almost all of us do, in fact, even if we're not necessarily opposed to the establishment.

But in almost all cases we do so because we don't see a real alternative and, after all, money has always been issued by the government, right? Therefore, it must be real. Bitcoin, on the other hand, is surely only a large scale global experiment.

There is some truth in this. After all, the Bitcoin experiment is only twelve years old.

But here's the thing. The money we know is also young. In fact, very young in terms of human history.

At 49 years old, fiat itself can still only be classed as an experiment. An experiment that, although widely adopted, we already know cannot work as a long term solution.

Without something else in place, that could be disastrous for economies of the future.

The Fiat Experiment

In broad and slightly simplified terms, currencies came off the gold standard in 1971 and became truly 'fiat' versions of their previous selves. Governments were truly free, in relative terms, to explore their own monetary policy while keeping an eye on what their actions did to international values and domestic inflation rates.

Of course, on a macro level, pretty much everything is theoretical anyway.

For example, we can be fairly sure that raising interest rates will lower certain types of inflation over time, but none of this is absolute and the sheer number of variables is impossible to calculate.

We're also pretty certain that printing lots of fiat money will lead to devaluation, but we have no idea of the timescale or to what degree. Does it happen all at once? Or gradually then suddenly? History gives us multiple examples but without providing explanations as to why one version may be different from another.

But the main point is that I am actually older than the money we use today. The fact that I can, figuratively speaking, look at an entire global monetary system and say "I am older than you" is simply an astonishing thought.

We are still in the experimental stages of fiat. We're still learning what the limits are and what levers affect what. We are also beginning to understand, through impossible and unsustainable global debt, that this simply will not work in the long term.

So where do we go from here?

The Bitcoin Experiment

Bitcoin is only 37 years younger than our entire fiat system, not even a blink of the eye in human terms.

We can already be certain that the fiat experiment will fail, or, at the very least, have to be significantly modified to continue to operate as-is. The only thing we don't know for sure is the timescale. It could be next year, it could be ten years, although it's unlikely to be more than a generation away using any basic extrapolation of current numbers and trends. Put simply, beyond that point, none of the numbers add up.

However, Bitcoin's numbers, again using basic extrapolation, provide an exactly contradictory picture over the same timescale. In basic terms, as fiat devalues Bitcoin holds steady with a predictability that can be calculated to eight decimal places. More, if, for some reason, you wanted to go that far.

So, is Bitcoin simply the latest replacement? Perhaps, perhaps not, but it is definitely something that hasn't been tried before, and for several solid reasons. One, however, is of particular importance on a global level.

Until now, the key point of failure of any monetary system has been a human intervention. We just cannot resist either tinkering with it, printing more for our own gain (either collectively or individually), or finding some other way to abuse and ultimately destroy it. Put simply, we can't be trusted with it.

Bitcoin, either by design or as a useful by-product of the design, recognizes this human failing and eradicates it as a point of failure by effectively removing us from the equation completely.

This is entirely unprecedented in human history. Money has always been controlled, influenced and ultimately abused by the few in power. We have never had a medium where this has been impossible. This is truly uncharted territory.

The theory says this should be a strong advantage, but we won't know in practice whether that is true for a very long time. After all Bitcoin, like fiat, is an experiment.

The Technology of Money

Economists will tell you that money must have three main attributes for it to be truly considered so. These are:

- It has to be a medium of exchange
- It has to be a unit of account
- It has to be a store of value

All forms of money have fulfilled these requirements to different degrees, but on a broader level, money can, in fact, be anything you and I agree it is. Literally, anything. Our mutual agreement can trump pure economic definition, at least on a micro-level.

But more importantly, the money we have exchanged between ourselves on a macro level has actually always been about the latest technology fulfilling these attributes.

Precious metals replaced barter, regulated coins replaced irregular precious metals, notes replaced coins (in terms of precious metal content) and digital representation replaced notes and coins to the point where the vast majority of transactions are no longer carried out in cash.

Over thousands of years of human history, we have continuously improved our monetary systems, each time experimenting with new approaches, better transmission, higher store of value by weight, levels of scarcity and control, and so on.

Most of the time we have improved it, learning from mistakes of the past, but not always. Some experiments failed. Some we have tested to destruction. All have been centralized.

In this context, Bitcoin is an entirely natural step. Bitcoin is the logical conclusion of the coming together of technology and thousands of years of evolution.

It is simply another experiment in a long line of never-ending experiments.

So, next time someone points out that Bitcoin is just an experiment, you can wholeheartedly agree.

Because it is.

Just like that cash in your pocket.

Bitcoin is the Perfect Asset in this New Investment Paradigm

33

Written by Mark Helfman

In the spring of 2020, the global financial system almost imploded.

In response, the world's governments flooded the markets with money and cut interest rates.

Time will tell whether these massive, coordinated, unprecedented intervention saved the system or delayed its inevitable collapse, but one consequence is clear:

Central banks and governments destroyed the value of safe, low-risk assets.

In doing so, they created a new investment paradigm. Bitcoin is the perfect fit for this new paradigm.

Yield, Where Are Ye?

Where can your money earn any money anymore?

Cash gives you zero interest (and sometimes costs you money). Savings accounts do barely any better.

Most bonds can't cover the rate of inflation. Bonds that do cover inflation come with a high risk of default.

Corporations have slashed dividends. P/E ratios have skyrocketed. IPOs now fetch outlandish valuations they can't possibly sustain over the long term.

Gold offers no yield, you never know if it's real, and it's expensive to buy and sell.

When you account for risk, *every asset has a poor expected return on investment.*

True, if the dollar continues to fall, you can make good risk-adjusted returns in emerging market debt and commodity markets. And investment properties always deliver great returns if you know how to manage real estate.

But few people have access to those investments or know how to use them. Also, I'm not sure you ever want to bet against the U.S. dollar over the long term.

All Dressed Up and Nowhere to Grow

On the flip side, U.S. investment firms have $26 trillion in assets under management, and, based on U.S. Federal Reserve data, U.S.

households have more than $19 trillion in cash, savings, and money market accounts.

A similar situation exists in other countries, too, albeit on a smaller scale.

On top of that, global funds, foreign investors, and offshore accounts hold upwards of $40 trillion. Nobody knows how much sits in corporate treasuries and business reserves.

That money wants to grow, but investors have no good place to put it.

So far, that money has mostly NOT gone into bitcoin. Of the $10 trillion in new money governments printed earlier this year, less than .2% ended up in bitcoin investment funds and personal wallets.

Guess what's changed since then?

Nothing.

Except the price has gone up for a while.

And That Has Made All the Difference

For months, bitcoin's returns have far outpaced those of every other asset class.

No wonder institutional investors, rich people, and a few corporations have started putting some money into it. Nobody has the patience and discipline to continue losing money with cash, stocks, gold, and bonds while bitcoin's price explodes.

At some point, once altcoins get large and liquid enough for big money to buy in any substantial amount, these same investors will add some exposure to them.

It's amazing how people's perspectives change when prices go up.

While Google Searches for Cryptocurrency suggest normal people haven't noticed yet, that will change soon, too. Rich people and insiders have made it seem safe, entrepreneurs have made it easy to access, governments have made it regulated and legitimate, and bitcoin's price has made new highs.

We have all the ingredients for a massive boom.

Traditional assets have more risks and less upside than ever before, while cryptocurrency has *fewer* risks and *more* upside than ever before.

Where do you think people will put their money?

Wrong! Fraud, Ponzi, Rat Poison Squared!

Ok, good point. In the short-run, *any asset's price can go up*, sometimes for much longer than you expect.

That doesn't mean it's a good investment or a good time to buy.

What about over 12 years? Is that long enough for an asset's price to go up before you consider it a decent investment?

How many people need to build businesses on and around bitcoin before you take it seriously? How high does price need to go before you feel like you're missing out on a big thing?

No "Real" Investor Thinks That Way

True, but most people are not real investors.

Real investors always have an investment thesis, a fact-based reason to expect prices will go up in the future.

For example, my investment thesis for bitcoin is simple: price always goes up and it works when the financial system doesn't. Factually correct, easy to prove, easy to understand.

If that seems too simple, you're probably right. When you have compelling evidence, facts, data, and history on an asset and its technology, you don't need to complicate things. I have 12 years of evidence backing my thesis.

So what? If everybody makes the same decisions as real investors, you can expect the same result. Money is money.

They don't need to have a thesis. They just need to act as they do.

(And they will.)

But COVID-19, Lockdowns, and Socialists!

People forget that the global economic recovery is fragile. In some countries—e.g., the U.S.—government support programs will expire soon, potentially sending the economy reeling. Sure, we have a COVID-19 vaccine but probably 1/3 of people will refuse to take it. Even if they do, we have a long way to go before we get back to normal.

Unemployment remains higher than normal and businesses remain strained.

Meanwhile, lockdowns make like unbearable for too many. And, some worry socialists will takeover government, destroy the world's economies, and impose draconian, liberty-crushing regulations.

While that's really bad, none of it seems to matter for crypto.

This spring, during global lockdowns and civil unrest, the entire crypto market doubled in size, and big money gobbled up bitcoins by the ton.

Now, prices continue to rise *despite* COVID-19 cases and deaths going up, and economic growth stalling.

Facts Don't Lie

Meanwhile, we have seen bitcoin repeat the same patterns and correlations we've seen for 12 years, regardless of the global financial, economic, or political situation at any given time. I have covered this extensively in my newsletter, *Crypto is Easy.*

Researchers from Fidelity, Morgan Stanley, Yale, and smaller institutions have shown a small allocation to bitcoin yields better returns with less portfolio risk. Bitcoin still keeps making new highs, with the usual pullbacks along the way.

We can't predict the future, but we can use this data to prepare for the most realistic outcomes. For bitcoin, that outcome is clear:

Up.

See With Your Mind, Not Your Eyes

In the short run, anything can happen.

In the long run, certain patterns and trends always play out the same way. Often, they're so compelling, you can't deny them—even though others will.

Today, we have the highest bitcoin price in years, continuous growth in altcoins (prices and technology), a positive macroeconomic environment, a traditional financial system increasingly unable to deliver the returns people expect, tons of money sitting in bank accounts, and underperforming investment funds, and a COVID-19 vaccine (with more coming by the end of this year).

Fundamentally, *nothing* has changed with bitcoin in the past six months. Same risks, same technology, same opportunities. It sure feels different, though—doesn't it?

It's amazing how people's perspectives change when prices go up.

What do you think will happen when its price goes higher?

Bitcoin is Worth $1,000,000. Now What?

34

Written by Owen Fernau

Recently, a tweet caught my eye about the potential for Bitcoin to become less secure as it scales.

Justin Bons:

"BTC has to double in value every four years for the next century or sustain extremely high fees

Such growth is impossible since it would exceed global GDP

Fees will never reach sustained extremes, due to the ratcheting effect of the fee market

Therefore; BTC security is doomed"

I jumped a bit. The idea seemed kind of correct. Everyone's talking about $100,000 Bitcoin and there are rumblings of a $1,000,000 price around the mid-2020s. The stock-to-flow model corroborates this potential increase.

Still, Bitcoin isn't universally accepted to be worth a million dollars a coin. If it was, it would already be there. I dug into the idea that Bitcoin's security will deteriorate as the project continues to mature.

I searched for more in-depth criticisms of Bitcoin's inability to scale (while remaining secure). The concerns appeared valid, and, for BTC investors, worth worrying about.

Before diving in, a basic idea to remember is that Bitcoin miners are compensated through two avenues: the block reward and the transaction fees.

At present, around 88% of miner revenue comes through the block rewards, while the remaining 12% is made up of transaction fees.

There are three major forces at play affecting these two percentages: Bitcoin halving, Bitcoin's price, and the transaction fee.

The Halving

Bitcoin's halving, which occurs approximately every four years, both hurts and aids miner revenue. It hurts miner revenue in that the block rewards are cut in half. It helps miners because the rate at which the currency enters the system has been halved too. If demand remains constant, Bitcoin's price will rise.

Halvings do hurt miners' revenue in the short run. Perhaps unsurprisingly daily miner revenue was cut in half in the days following the 2016 and 2020 halvings. Though within a year after

2016's cutdown revenue was back to pre halving levels (due to Bitcoin's rise price along with increasing demand for block space).

In 2020 miner revenue is already up to July's pre-halving levels.

Over the long term, halvings further the march towards a transaction fee dominated miner revenue model.

The Price of BTC

A heightening price of Bitcoin helps offset decreased miner revenue from halvings. If mining revenue consisted only of block rewards, then Bitcoin's price would need to continue to double forever, in order to maintain its current security budget.

(As a side, the "security budget" essentially means the miners' revenue. The miners keep the network secure and are paid to do so. If the system becomes corrupted, meaning one party gains control of over 50% of the mining power, that group can rewrite Bitcoin's history and hence destroy the cryptocurrency's legitimacy.)

But miner revenue is not solely composed of block rewards. There are transaction fees too, which allow for the possibility of stable miner revenue with a flat Bitcoin price (despite the halvings), offset by transaction fees.

The Transaction Fee

Bitcoin blocks can be thought of as real estate. People pay for space in the block in order to send a transaction. And that payment constitutes the second source of revenue for miners.

Transaction fees are meant to constitute an increasingly higher portion of miners' total revenue, and this must happen (if the system is to remain secure, and hence useful) as Bitcoin's price cannot keep doubling ad infinitum to compensate for the halving block rewards.

What does a post-$1,000,000 Bitcoin world look like?

So, if Bitcoin's security budget has to remain the same, but Bitcoin's price has leveled off at around a million dollars, transaction fees would need to increase to offset the halvenings.

Is this true?

A hole in this sort of thinking is that the Bitcoin security budget needs to be as high as it is. Bitcoin's security budget settles arbitrarily at the level it does. No one in the Bitcoin ecosystem is primarily concerned with the blockchain's security budget. Miners care about it because of revenue, but not primarily because of the security it provides.

This is not to say Bitcoin doesn't need security spend. It does. But whether it needs to maintain its current level (a highly variable $18 million a day), isn't certain.

What's going to happen?

Regardless it seems like the crux of the issue about Bitcoin's security is whether users will go elsewhere once transaction fees hit a certain height. Or whether miners will shut down as transaction fees fail to offset halving costs as Bitcoin's price stabilizes.

It remains to be seen. In order to survive though, Bitcoin must continue to be seen as 'prime real estate' on which to transact. There are many blockchains. If an equally secure blockchain emerges, with lower transaction fees, users may leave Bitcoin. Fewer users means lower transaction fees, means lower security, means Bitcoin loses its valuable property status. As transaction fees gain in importance for the system it's vital that people still wish to transact in Bitcoin (for the original cryptocurrency to survive).

Closing

As Bitcoin's price levels off, prudent investors will monitor Bitcoin miner revenue composition to see whether transaction fees are really going to pick up the slack where the block rewards leave off.

Luckily, if the stock-to-flow model is any good, investors have some time to ride the wave, before it potentially crashes.

When Will Bitcoin Reach Critical Mass?

35

Written by Thomas Daniel

In physics, critical mass is defined as the minimum amount of material needed to maintain the nuclear chain reaction. What we're watching, what we've been watching in the cryptocurrency markets, is an attempt by Bitcoin to reach critical mass and become that digital medium of exchange the world seems to want it to be. When investors are wondering, will it achieve the massive growth we're all expecting?

Bitcoin has been in a bull market all year, despite some of the roughest conditions in history, and it has held its ground remarkably well. Institutional interest has increased astronomically this year as the government has opened up avenues for cryptocurrency to enter the mainstream. Everything from savings to digital ETFs to PayPal is getting involved in cryptocurrency because it is faster and more secure than other transaction verification methods.

The stimulus package currently under debate in Washington is a major potential influence to watch for on the bullish side, but Representative Maxine Waters and a small coalition of Democrats are working to reign in what they see as a potentially confusing and therefore manipulative market. In the present administration, Trump has said repeatedly that he doesn't like cryptocurrency and we haven't seen much major legislative support, as we might from proactive Democrats who want to tighten up the market.

The New Bear Thesis: Government Intervention

Despite the unlikeliness of the scenario in which Democrats ascend to control of the federal government and decide to shut down cryptocurrency markets, it is nonetheless an unpredictable situation. Regulation is on the table because it is needed to help eliminate fraud and overt manipulation in the crypto markets, and this stabilizing influence is something that would ultimately send prices higher over the long term.

Despite the obvious need for a more crypto-friendly regulatory environment, the newly minted bear thesis posits that legislation could be counterproductive somehow. Pelosi faced a primary challenge in 2019 that forced her to understand some of the cryptocurrency issues of our time, but the Trump administration's laxity about who should be able to receive a bank charter is undeniably problematic. The same article points out that self-hosted wallet regulation is unpopular, and ultimately this all adds up to the same final tally: the US is in for a complex legislative reckoning on the issue of cryptocurrencies, and it might be handled improperly.

Still, it's hard to see this as an entirely bearish situation, with Janet Yellen tapped for Treasury Secretary. The fact of the matter is simply this: if institutional players get involved, and they will have to for Bitcoin to continue to grow, regulation is a necessity and not just an inconvenience. Institutional money eschews unregulated markets.

The Enduring Bull Thesis: There's No Stopping Us

As challenging as times are, the case can still be made with relative ease that none of these developments on the regulatory side of the situation here will be enough to undo the newfound institutional interest in cryptocurrencies. For one thing, crypto is over $500B in market capitalization, still. As that number approaches $1T, as it is likely to over the next few months, if not within the next month, it will become more difficult to do anything with legislation that would jeopardize the function of the networks. It isn't impossible, just far less likely than ever before.

Additionally, cryptocurrencies are largely decentralized. It is possible that crypto could run afoul of some sort of regulation and *still remain a viable medium of exchange.* These larger conflicts are likely to remain phantasms from far away, however, so there is little chance of the ultimate showdown ever taking place. Instead, what is most likely is that the legislature will listen to the experts and take an active interest in allowing cryptocurrency to move ahead.

Critical Mass For Bitcoin: A Popular Consensus

The funny thing about cryptocurrencies requiring consensus to function is that, more broadly, there is little in the way of consensus at all. We have the Bitcoin Bull camp on the one hand, which seems to be growing, and on the other hand, we have all of the skeptics and the folks who haven't looked into it and everyone who thinks it's just too complicated.

The technology is becoming more accessible fairly rapidly, but it is difficult to say when it will have done enough to become the popular medium of exchange every cryptocurrency asset ultimately seeks to be. Another media blitz a la 2017 does not seem to be forthcoming, despite the recent setting of a new all-time high price for Bitcoin, but prices remain very high and so does the potential for further appreciation.

The oft-delayed stimulus package is being negotiated again, with December 11 set as a hard deadline because the government runs out of money if a new spending bill is not approved. Pelosi and McConnell are more incentivized to make a deal now than they've ever been before, and it seems highly likely that some sort of additional unemployment payment will be rolling out in the next few weeks as a result. In the past, we've seen the Bitcoin price surge following stimulus as people struggle to save and earn appreciation on any spare money they can save up during these trying times.

Could this be the push Bitcoin will need to break through the $20k psychological price barrier and run higher? Time will tell, of course, but I for one am long BTC on the possibilities here. Additional institutional interest will probably also push the needle where it needs to go. There are too many ways for things to work out in favor of cryptocurrencies today and not enough ways for them to go wrong.

When Bitcoin hits critical mass, the price will skyrocket and then stabilize at a very high number. The thing is, that high price will be hard to move because there will be so much invested in arbitrage from any angle that people won't be able to push it around much.

And that's when Bitcoin goes from speculative asset to global reserve currency.

Cryptocurrency is Not as Big as You Think . . . Yet

36

Written by Mark Helfman

With bitcoin hitting all-time highs for the first time in years, you may think it's getting big. Huge, even.

Wall Street luminaries like Paul Tudor Jones and Jamie Dimon are telling people to buy it. Long-time skeptics like Nouriel Roubini begrudgingly admit it's a "partial store value."

S&P Dow Jones announced it will create crypto indices in 2021. Former bitcoin-basher JP Morgan is now building a book of business among Crypto companies. Goldman Sachs formed a digital assets group. Fidelity has had one for almost two years.

You ain't seen nothing yet.

A Thimble's Worth of Water in an Ocean Full of Wealth

Sometimes we look at crypto and see big numbers.

For example, $300 billion in bitcoin, $16 billion in stablecoins, $6 billion in Graystone Bitcoin Trust, $13 billion assets on DeFi lending platforms, 75 million bitcoin wallets, 15 million active daily users of Brave browser, and so on.

Once you start talking about millions and billions, it seems like the market is huge.

For us, maybe. For the world, it's not. Look at the numbers:

- More than $19 trillion in cash, savings, and money market accounts for U.S. households alone. When you include savings from people in other countries, that amount goes much higher.

- Over $26 trillion in U.S.-registered investment funds.

- More than $100 trillion in government debt.

- Over $200 trillion in corporate and household debt.

- At least $400 trillion in financial products like derivatives and collateralized loan obligations.

- Hundreds of trillions of dollars in real assets like property, cars, and collectibles.

In total, over $1 quadrillion worth of "stuff" can get recorded on a blockchain and exchanged using cryptocurrency.

If a $20,000 bitcoin seems crazy and all these altcoins seem overpriced, consider the bigger picture.

So Many Ways to Capture Wealth

What makes me think cryptocurrency will capture any of that wealth?

Because traditional commerce can't compete with global, open, permissionless financial networks that record and track the movements of things with mathematical precision and no middlemen. It costs too much to maintain.

In contrast, blockchain technology continues to scale and innovate. Some governments have already moved to digital currencies. Banks have quietly pocketed hundreds of blockchain patents.

As everything moves to a natively digital environment, *something* will capture its value—even if only as stablecoins, synthetic representations of real-world currencies.

Developers have already developed protocols that give users all the benefits of traditional finance with no commissions and a share of the platform's success.

As a result, they're quietly building decentralized margin accounts, insurance funds, staking rewards, synthetic assets, and liquidity pools for exchange assets and derivatives—with almost all of the benefits going to those who use the platform, rather than a broker, custodian, or investment firm.

Over the long run, which service will win: the one that costs almost nothing and returns the benefits to users, or the one users have to pay for and keeps all the benefits for itself?

Think about token sets and PIES, which offer non-custodial access to diversified investment portfolios. Rather than pay somebody for access to investment funds, you buy a stake in a protocol that does all the work for you.

Or, consider applications that maximize yield by moving money from one platform to another. When you can move unlimited amounts of value at virtually no cost, you probably will.

The catch: in order to do these things, you need some way to capture that value in a form that people can freely exchange with each other. In other words, you need cryptocurrency.

When you anticipate everything will run on a blockchain within a few decades, $500 billion doesn't seem like much.

Not Just DeFi

And what about non-financial markets? Cryptocurrency technology is widely applicable.

The world wastes a lot of energy. Some places produce more than they need while other places face chronic shortages.

Once you tokenize energy consumption as a cryptocurrency, you can create secondary markets and new distribution channels. You can embed carbon emissions credits, rebates, and incentives into the tokens. Power plants and distributors can mint their own tokens when they create energy and burn those tokens as people consume them.

What about decentralized autonomous organizations for corporate and community assets? Security tokens for real estate development?

With cryptocurrency, you have seemingly infinite ways to capture wealth. It's just a matter of building the right platforms.

Fortunately, we have a lot of people building those platforms. They toil in the shadow of bitcoin and DeFi, but their work will get its proper recognition. They've barely scratched the surface of what they can do.

Slowly, Then All at Once

If that seems like the pipe dream of some distant future, it's a dream that's getting built today. Once it's ready, you'll be surprised by how quickly it becomes a reality.

Until June, you probably hadn't heard about DeFi. Now, it's a big deal.

Did you know that some of these DeFi projects spent three or more years in development?

When I started working on *Consensusland*, I consulted with a few projects, including Aave. At that time, they called themselves ETHLend. They had barely gotten started.

Who would have known they'd emerge as a top DeFi project two years later?

It takes a long time to become an overnight sensation. Once it happens, everything changes.

Despite bitcoin hitting its previous all-time high, crypto is still a very small market. How big could it grow?

The moon's the limit.

Inside Gitcoin: The Project Boosting Open-Source Builders Around Crypto

37

The cryptoeconomy had its share of bright stars in 2020, and among these amazing projects' ranks was Gitcoin.

Why Gitcoin? Because the entire cryptocurrency ecosystem is founded on open-source software development, yet OSS builders have historically had a difficult time acquiring adequate funding. That's where Gitcoin comes in.

A decentralized platform atop Ethereum, Gitcoin is a community infrastructure that lets folks working in and around crypto OSS readily monetize their labor like never before.

To help these builders earn their keep, Gitcoin facilitates gig bounties, hackathons, as well as the project's flagship Gitcoin Grants service. This increasingly popular service sees sponsors donate money into Gitcoin's Quadratic Funding (QF) Fund, which uses a special QF algorithm to match community donations to OSS projects.

This novel funding style makes it so that even a $1 donation to a project can fetch another +$100 in matched funds to that same project!

Example: My DeFi Arts Newsletter is listed in the ongoing Gitcoin Grants Round 8. A donation of 1 DAI (i.e. ~$1) generates me a further 8 Dai in matched contributions presently.

As the Gitcoin team explained of its QF system last month, the program is all about helping projects become self-sufficient in grassroots fashion:

"If you're a grantee, the key thing to note is that the number & amount of each contribution to your project influences the total amount allocated to you. If you're a contributor, even a small contribution can have a huge impact on the amount of matching funding a grantee gets. [...] In other words, Gitcoin Grants is really all about communities working together to help fund and signal support for public goods. And we all benefit from public goods."

The ecosystem is really starting to rally around this model, too. The matching pool for Gitcoin Grants Round 8 is $500,000 USD, the largest pool yet. That funding was raised from some of the cryptoeconomy's biggest names, too, like Binance and Chainlink.

Not Exclusive to Ethereum, Either

Gitcoin isn't just a force for good on Ethereum. This year has notably seen Gitcoin enter into and back a series of partnerships, hackathons, and bug bounty programs with non-Ethereum projects. These projects have included:

- ETC
- Sia
- Cosmos
- NEAR Protocol
- Polkadot
- Zilliqa
- Arweave
- Matic Network
- Chainlink
- Filecoin
- Zcash

That said, look for Gitcoin to keep making inroads with other prominent blockchain communities as the project continues to gun to become the wider cryptoeconomy's premier OSS dev hub for years to come.

That's good news for the crypto space because the first 7 Gitcoin Grants rounds already fetched a few million dollars' worth of OSS funding. This reality suggests this grant system can become much larger and much more of a productive tour de force as Gitcoin and the cryptoeconomy grow.

In the meantime, it'll be interesting to see how Round 8 plays out. Donors are already off to a hot start and with this round's matching pool being the largest one yet, it seems like Gitcoin itself is only just beginning to heat up.

Bitcoin Made Wikileaks Resilient

38

Written by Mark Bailey

Publishing American state secrets put Wikileaks at odds with the US government. Wikileaks was financially blockaded. The organization established a Bitcoin payment channel on June 14, 2011. This payment channel allowed Wikileaks to persist despite being blacklisted by the legacy economy.

The politics of the Wikileaks situation are arguable. The organization published footage showing US military personnel murdering civilians and journalists in Iraq. That seems like the kind of thing that people should know about. But I couldn't even guess at the national security implications of leaked military data. And Julian Assange, Wikileaks' founder, is the definition of a controversial figure.

Politics aside, Wikileaks is an interesting Bitcoin success story. The leak publisher was under intense financial pressure. "American leadership reached out to choke any point of support for the organization," *according to Forbes*. "The Swiss bank PostFinance froze the personal account that belonged to Julian Assange. MasterCard and Visa refused to process any more transactions for Wikileaks." Despite this pressure, Wikileaks was able to operate for years thanks to Bitcoin. In other words, Bitcoin made Wikileaks resilient.

Because Bitcoin was decentralized, central authorities could not use their political might to cut off Wikileaks' funding. Nor could the intelligence services of major governments stop the organization from using Bitcoin. Blockchain technology is inherently resilient. But Wikileaks proved Bitcoin's resilience in an extreme, real-world situation.

Remittances

Wikileaks was financially blockaded for political reasons. But many people in the world are blocked from using financial infrastructure purely by the absence of proximity to it. Cryptocurrencies are contributing to the development of a more accessible alternative infrastructure. This alternative makes the communities it serves more resilient.

One important area where crypto is helping people to skirt de facto financial blockades is international remittances. The people sending these remittances are often migrant workers sending money back to families in developing nations. Globally, international remittances cost an average of about 7 percent of the total sum sent. According to Reuters, if this cost were reduced to 3 percent, those sending this money would save $25 billion annually.

It is possible to send cryptocurrency to anyone in the world for less than 3 percent. Given the negligible cost of electronically transferring funds, companies charging 7 percent for this service appear predatory, and companies charging 3 percent appear just a little less predatory. Crypto is changing this game. According to an article on the United Nations' website, "cryptocurrency-based remittance services are popping up in various countries. These services include Abra, which operates in Malawi and Morocco, GeoPay in South Africa, BitMari in Zimbabwe, and London-based Kobocoin, which was launched by Nigerian entrepreneur Felix Onyemechi Ugoji."

One-Woman Wikileaks

Maltese investigative journalist Daphne Caruana Galizia was assassinated by a car bomb on October 16, 2017. Although she was described by *Politico* as a "one-woman WikiLeaks," this anti-corruption journalist did not share Wikileaks' enthusiasm for Bitcoin. "I suddenly [realized] that here we have a situation in which the governor of the central bank of a European Union member state, and its prime minister, are arguing for the rise, growth, and promotion of crypto-currencies, which will give enormous advantages to money-launderers, terrorists and other forms of [organized] crime," Caruana Galizia wrote in a blog post. "Whoever has a vested interest in the heightened use of cryptocurrencies is already [funneling] money into the Labour Party and a couple of Panama companies, and calling the shots."

In the aftermath of Caruana Galizia's death, many Maltese politicians that had been supporting the rise of crypto resigned. Political support for Malta's status as "Blockchain Island" evaporated. But in a way, Caruana Galizia's anti-crypto stance actually bolstered the case for crypto. By arguing against cryptocurrency on the grounds that it made criminals better at skirting centralized authorities, Caruana Galizia was also arguing that crypto makes its users more resilient.

Indiscriminate Resilience

Julian Assange once called Bitcoin the "real Occupy Wall Street." Maybe there's something to that. Cryptocurrencies subvert a status quo where too many people are held hostage by the legacy economy. Bitcoin rendered the legacy system powerless against Wikileaks. New crypto companies are challenging the predatory practices of the international remittance industry. Corrupt politicians even seem to be getting in on the game after discovering that they, too, can become more resilient by embracing crypto.

These three examples -- Wikileaks, international remittances, and organized crime -- illustrate that crypto is already an unstoppable force. Cryptocurrency confers resilience indiscriminately, which is as it should be. While this may mean that central authorities have to develop new methods of curtailing crime, it also means despotic regimes have fewer tools of oppression at their disposal. And that average people have financial options they never had before

Can We Get Rid of Fake News and Trust the Web Again?

39

Written by Jillian Godsil

In June 2020 WordProof and Sebastiaan van der Lans were awarded the top prize of Blockchain for Social Good by the European Commission and with it a whopping €1 million. They beat off 175 contenders from more than 40 countries at various rounds to emerge the overall winner with a voting score of 29.5 out of 30 by the voting jury. So what is the WordProof project and why did it win?

Sebastiaan starts by explaining that the internet as we know it now is broken. He maintains there is a deep-rooted issue with its trustworthiness. Currently, 29% of Europeans are suspicious of the internet with 86% of Europeans saying they have fallen for fake news online.

"Here at WordProof, we want to bring back trustworthiness so that what I read is real and I can verify who wrote it. And then in the example of an eCommerce dispute, how can we retrace the steps? Well, our solution was to leverage blockchain timestamps so we can return to a trusted web – for everyone."

WordProof timestamps content on the blockchain which is helpful for viewers but also for search engines. If changes are made after the original posting, then these are highlighted too. Perhaps the revisions are helpful like a rolling news item, but sometimes they might be unhelpful.

To give a real example, we could look at the recent UK political fallout in April during their lockdown. The US chief advisor to the prime minister, Dominic Cummins, made headlines beyond his breaking of the lockdown when changes were spotted on his blog from the previous year to appear to make him more expert if not actually prescient on the issue of the coronavirus.

According to the Guardian newspaper, a data scientist spotted the new edits using periodic snapshots of Cumming's blog saved by the Internet Archive. But these changes were only spotted as a result of intense scrutiny of his movements during the political furor. Otherwise, the insertion might not have been spotted and the advisor might be able to claim he was much more knowledgeable than he actually was.

WordProof does this automatically to all WordPress websites when installed. Given that some 38% of all websites are based on

WordPress this has the capacity to radically update the content served online.

"Exactly, but even on rolling news sites, it would be good to see what was there before. Has an editor radically changed a viewpoint, added in beneficial news, deleted controversial details. The revisions can be seen – very clearly what was added and more importantly what might have been taken away."

In simple terms, text altered is color-coded; with green text showing additions and red showing what has been removed. It's like track changes only time-sensitive.

WordProof was launched at WordCamp Europe, the largest WordPress event in the world where Sebastiaan gave a keynote speech. And it has really caught on with hundreds of sites signed up and more than 250,000 pieces of real information timestamped, which in turn have been served on the internet more than 100 million times.

"It's not just about the technology it's about social good, making the internet safe for grannies, grandchildren, and everyone in between. It is also about rethinking making mistakes."

Sebastiaan notes that there is a kind of taboo about making mistakes. People lawyer up and become aggressive rather than just putting up their hand to acknowledge the mistake.

"Wouldn't it be better if news providers can say we want to be transparent and if that means showing our mistakes, so be it. And we need consumers of news to say it's okay to make mistakes."

It's a good point but it will need a seismic change in society and issues like defamation and libel will need to be re-evaluated. If an editor or publishing house gets it wrong, will it be enough for the paper to apologize and correct the mistake or will they still be sued for damages? Will an apology – with transparency – be enough?

Sebastiaan argues that building a trusted web is also going to take some leaps of faith. "Transparency is key as well as linking identities to the content. Combined - they offer accountability and can be ranked higher in a trust score – surely that is to be welcomed by sincere content providers?"

Sebastiaan has been involved in the WordPress community for more than 13 years. He is also an early ambassador for the Europechain project based out of the Netherlands. Europechain is a GDPR chain offering enterprise-level solutions for its customers. While blockchain is agnostic, he loves what Europechain is doing for GDPR and the protection of people's information.

"It'll work when corporations get on board, and that is what Europechain is doing."

The more people working to make the interest a trusted world again, the better.

Despite a +130% Premium, Institutions Can't Help but Buy Ethereum (ETH)

40

Written Dalmas Ngetich

As they say, a person's perspective of life depends on where they are standing. Such differing views mean that two people could have a different understanding of the same thing.

To some, Ethereum is an experience; an actualized decentralized innovative platform that unleashes the power of the blockchain.

To others, it is a speculative asset. They view it as an opportunity to make money, and in their wisdom, they have decided to bet on Ethereum's future.

In the last three years, right after the hysteria of late 2017, the winter of 2018 crashed hopes. Regardless, Ethereum has shown thick skin in bouncing back better and stronger.

Roughly halfway through the journey to 2017 peaks, there are interesting metrics that are being observed by trackers.

- The hash rate is up. Blame it on prices.
- The transaction count is higher than Bitcoin. This is largely on account of DeFi, which has been a beast this year.
- Gas limits and fees are up. Blame it on DeFi

And lastly,

- Volume in transactions over $100k is at 2017 levels.

Interesting but not shocking.

Ethereum is more than just a transactional layer. It is a smart contracting platform with tones of moving parts. At the time of writing, there is lots of development going on. Refinement of all forms.

For instance, there are two chains running concurrently, leeching value from the same native currency. While at it there are layer-2 solutions.

CryptoWriter has exhaustively covered these two topics. Read more about these interesting developments below:

- Vitalik Proposes a Change in Ethereum's Roadmap to Prioritize Rollups

- Ethereum Layer-2 Scaling Strategies, the Progress So Far

Going back to the basics and eliminating the smart contracting bit of Ethereum, it is interesting to note that the network is increasingly being used for bulk transactions.

Pound to pound, there are alternative platforms that may replace Ethereum if expediency matters. Transaction fees are quite high—actually rising over time as the network's activity heats up, but still, users seem not to care.

The only explanation for this is institutional influx.

Heavyweights are re-assessing their positions; digging deeper and realizing that Ethereum (and blockchain) isn't a casino stuffed with scams as they thought. ICOs have fizzled out, and law enforcement did their job in arresting perpetrators. It has since evolved, and public funding is now more open, investors more demanding, and laws are better than they were in 2018.

So dense is Ethereum now that the main objective is to address impediments presented first by scalability, which influences transaction fees.

Developers are taking a two-pronged approach to this matter. First, layer-2 solutions where Optimistic Rollups as a viable option is pulling ahead, and the multi-stage, multi-year Eth2 culminating in Serenity introducing shards.

When both are live, they will have a multiplicative effect—they won't be competing solutions, further making it faster, secure, and cheaper to exploit DeFi yields, play games, tokenize assets, move value—like Bitcoin, and so much more.

Coupled with supportive comments from regulators; comments from the US Commodity Futures Trading Commission (CFTC) Heath Tarbert fueled optimism. He said he is impressed with Ethereum and thinks DeFi will play a critical role in the future preventing systemic risks posed by megafunds as a revolutionary solution, ETH is drawing "serious" investors.

According to the Managing Director of Grayscale, Michael Sonnenshein, 2020 has seen a significant rise of "Ethereum only" investors. DeFi explains part of this but it is the dominance of the platform as a leader in smart contracting (amid Ethereum killers) and the preference from developers.

Like Bitcoin, institutions are now training their eyes on Ethereum and ramping up their purchase of Ethereum Trusts.

But here is the catch, for every share purchased—each trading at $118.70 as of writing (Dec 12), the investor owns a mere tenth of ETH at a huge premium of around 130 percent—down around 1,300 percent between Q1 and Q2 of this year.

Here it is: one ETHE Share==0.09303888 ETH

That means, for each ETH on the spot market trading at around $550 as of writing, Grayscale investors are buying each for over $1,100. This has been described as the cost of convenience and rigid regulations that can't keep up with the fluid and fast-changing events of the crypto sphere.

There are 25,933,100 ETHE shares outstanding attracting a 2.5 percent management fee. All in all, Grayscale is managing over $1.6 billion of investors' money dedicated to ETH.

From this alone, it is clear that investors are willing to pay more to earn ETH if it means doing it without risks and compliantly given the stringent US laws now that Grayscale Ethereum Trust reports to the SEC.

For the high premium, mechanisms are in place to eliminate what would be lucrative opportunities for arbitrage. Each ETHE share minted must be held for one year. And not just anyone can dabble with these trusts, there is a $25k hoop only for accredited investors.

Whether purchasers are doing so using 401k money or not, it goes on to show institutions (and retail investors) don't care about how far they go so long as they get a piece of the ETH pie.

Network effects might end up pushing ETH prices higher but prices—as it is, counts for nothing considering the platform's potential.

Is the Plunging U.S. Dollar Setting Us Up for the First Great Digital Gold Rush?

41

Written by Mark Helfman

As the value of the U.S. dollar continues to plummet from its March 2020 highs, cryptocurrency has found a new narrative: a hedge against the falling dollar.

While the potentially massive flow of dollars sounds great for anybody who wants bitcoin's price to go up, it means a whole lot of people will enter crypto for the first time, all at once.

That may be why Wells Fargo Bank warned its customers that cryptocurrency investing is like the 1850s gold rush, "more speculation than investment." It's not the first comparison.

Most notably, in 2019, at the depth of the great crypto bear market, aka "crypto winter," *Hacked* published an article also comparing bitcoin to the California gold rush, when the discovery of gold sparked a frenzied, greed-driven migration to California from 1848 to 1855 that left everybody poor except the frauds and scammers who ripped off speculators.

They're wrong. The gold rush hasn't happened yet.

When it does, we should all hope it turns out as the California gold rush did.

Bringing Wealth to Barren Lands

In popular history, people from all over the world went searching for gold in the hills around San Francisco and Sacramento. They found no gold and left with nothing.

Along the way, merchants, lenders, and corporations made fortunes off of their foolish hopes. The real wealth went to people with picks and shovels. The miners got nothing.

Guess what?

Popular history is hogwash.

By the time the gold rush ended in the mid-1850s, San Francisco had grown from a tiny outpost to a thriving city. At the time, it was considered the largest and most important financial center in the Western United States. Not much poverty, lots of growth. Sacramento became the capital of the fastest-growing state in the union. Almost everybody did well—including those first settlers.

Yes, those early miners suffered crazy-high costs for food and equipment. In 1849, a single egg could cost the equivalent of $25 in today's money, coffee went for more than $100 per pound, and a pair of boots could set you back more than $2,500.

By the gold rush's peak in 1852, prices had dropped a lot.

Maybe the influx of supplies through San Francisco tamped down prices? Perhaps later miners made less money than those early prospectors, therefore wouldn't shell out for luxuries like eggs and coffee? Could three years of investment in roads, ports, buildings, and laws have made it cheaper to get products to market?

When those early settlers arrived, the land was mostly empty hills and rivers. No ports, roads, or depots, nor any formal system of property rights, land grants, or taxation.

They brought steamboats, railroads, retail, and modern prosperity. Entrepreneurs discovered new dredging techniques and invented modern hydraulic mining. Agriculture boomed. Businesses flourished. Governance improved.

Within two years, California gained statehood. On top of that, some economic historians credit the gold rush for sustaining the longest recession-free period in U.S. history (1841-1856).

Can Bitcoin Do the Same Thing?

Perhaps cryptocurrency seems shady because most people remember 2017, when ICO scams, "pump and dump" investment schemes, and outright frauds ruled the industry.

Costs to buy and sell crypto reached 7% for some services, prices of every crypto shot up, and transaction fees rose to $20 for a simple bitcoin payment.

Then the whole market crashed. Three years later, it still hasn't recovered.

Probably not the best look for crypto, but most people don't realize how much the market has changed since then.

Anybody who invested during the bear market should be way up on their money.

If you haven't put any money into the market, you now have better payment rails and many low-cost, low-fee crypto service providers to help you get your money in.

Sidechains, advances in new blockchain technology, U/X improvements, and better regulations have made everything safer and easier. Fidelity, JP Morgan, and other financial institutions offer custody, security, and operational support for businesses dealing with crypto. Every major stock exchange operator has a subsidiary for crypto and professional financial firms now offer cryptocurrency funds.

Governments regulate the industry. Well-known investors and corporations invest in it. VCs own stakes in altcoin projects.

Developmental technology like Bitcoin's Lightning Network and Ethereum 2.0 can make crypto transactions faster, cheaper, and more secure than ever before.

Other altcoins have even more powerful technology. For example, permissionless financial protocols, hack-proof identity solutions, high-capacity data sharing networks, and platforms for creating unique, transferrable digital assets, and securities.

And they're all growing.

Prosperity Next?

Like the 49ers of the California gold rush, these new crypto pioneers are bringing money, talent, and infrastructure to a barren land.

Meanwhile, crypto-specific trade organizations have pushed governments to craft clearer regulations. Financial companies have created better infrastructure and traditional investors have started moving money into the markets.

Now, we're seeing new awareness, investment, and professionalism among people involved in this space. If you can believe it, the *Wall Street Journal* now covers bitcoin.

With all this positive activity, will we get the same prosperity as those old California miners?

The odds are looking pretty good.

Solving Poverty with a Crypto-UBI

42

Written by Mark Bailey

There is a problem in society that is often framed as inequality. Some people have very little while a few people have a great deal more. But framing this problem in such comparative terms can make it seem as if the underlying issue is fairness. From my perspective, the underlying issue is not fairness. It is poverty.

Whenever a basic need can't be met, there is poverty. Chilean economist Manfred Max-Neef identified twelve fundamental human needs: subsistence, protection, affection, understanding, participation, idleness, creation, identity, and freedom. At a minimum, life requires meeting the needs of subsistence and protection. So it is problematic if some members of society can't meet their subsistence and protection needs for financial reasons.

Small businesses are disappearing in the Covid economy. Workforce automation is producing technological unemployment. Food insecurity is way up. Poverty is or should be, a pressing concern. Universal Basic Income (UBI) is a basic income given to everybody. The introduction of a UBI could end poverty. And blockchain technology could facilitate a UBI in some important ways.

Basic Income Improves Wellbeing

In the US, poor people are twice as likely as their wealthy counterparts to die before reaching old age. Many factors, such as increased stress and limited access to medical services, contribute to this shocking statistic. Providing people with a basic income could mitigate these factors, increasing the longevity of the poor.

There is also evidence that a basic income could improve mental health. Recently, the National Bureau of Economic Research put out a working paper titled, "The Comparative Impact of Cash Transfers and a Psychotherapy Program on Psychological and Economic Well-being."

The paper's abstract states:

"We study the economic and psychological effects of a USD 1076 PPP unconditional cash transfer, a five-week psychotherapy program, and the combination of both interventions among 5,756 individuals in rural Kenya. One year after the interventions, cash transfer recipients had higher consumption, asset holdings, and revenue, as well as higher levels of psychological well-being than control households. In contrast, the psychotherapy program had no measurable effects on either psychological or economic outcomes, both for individuals with poor mental health at baseline and others. The effects of the combined treatment are similar to those of the cash transfer alone."

The implications of this study are important for mental health treatment. But they're more important for economic policy. They suggest that purely financial solutions can effectively address some of the mental health challenges faced by the poor. To me, that sounds like a strong argument in favor of a UBI.

Implementation

A recent article in the *Coin Telegraph* titled, "How Cryptocurrency Can Help in Paying Universal Basic Income" presents one vision for facilitating a UBI with crypto. Some of this article's claims are not ideal. It states, "[The state] carries out centralized cryptocurrency issuance, direct transfer of UBI payments to electronic wallets of citizens, control of cash flows within the system, and blocking of suspicious transactions, including the purchase of certain goods and services." This statement suggests that the article favors a Big Government approach to UBI implementation.

Blocking access to certain goods and services at the monetary level embellishes the UBI in an undesirable way. So does the idea of the state directly controlling cash flows. Also, making benefit payments selectively rather than universally reflects an unusual interpretation of universal basic income. But a UBI need not involve these things. It could shrink government, not enlarge it.

A primary benefit of establishing a UBI is the elimination of many complex, narrowly-focused government welfare programs. Instead of replacing these programs with a sprawling crypto nanny state like that *Coin Telegraph* article suggested, a UBI could be implemented in a simple and efficient way that reduces the need for government services.

In the US, Congress could have the Treasury Department direct the Federal Reserve to use IRS data to populate a new public blockchain with everybody's accounts. A new sector of the Government Accountability Office could be created to ensure that direct Digital Dollar payments to these accounts are properly administered. After a suitable time, redundant government programs could be cut.

For a UBI to be effective and sustainable, its assistance amounts would have to be tied to accurate inflation measures and adjusted continuously. To avoid payments just going to service debt, the establishment of a UBI would have to coincide with debt forgiveness. If Universal Adequate Insurance accompanied a UBI, the positive impact on people's lives would be enormous. All of this sounds great, but there's a catch, and it's a doozy.

The Cost

If 360 million Americans were each given a Universal Basic Income of $2,000 a month, it would cost $8.64 trillion annually. No combination of spending cuts and new taxes could make a government outlay of such magnitude appear palatable. US GDP only amounts to $20.5 trillion. Nothing about the current political climate suggests that there's an appetite for such spending in Washington.

At $8.64 trillion a year, a UBI would reconfigure the entire economy, necessitating a taxation scheme that many would likely loath. Yet the alternative to a UBI is a heartless status quo that manufactures poverty and leaves millions of poor people with insufficient means to meet their basic needs.

There are a few projects out there aiming to make a crypto-UBI a reality without government involvement. Some, like Zero Poverty, are just starting out. Others, like Circles and Good Dollar, already seem to be making payments to people. Although highly unlikely, it is conceivable that some combination of projects like these could eventually provide people with substantive basic incomes. And, regardless of their practical effectiveness, such projects serve the public good by raising awareness of the need for Universal Basic Income.

A Collection of Digital Art Pieces sold for $777,777, Will NFTs Flip Crypto?

43

Written by Dalmas Ngetich

$777,777 isn't any other number.

A prime number, true, but for NFT fans, magic.

It is the amount paid for a collection of digital pieces early this month, signaling the inevitability of NFT success and simply, the rapidity of the space's evolution.

In 2020, DeFi dominated, relegating other equally important projects to the back seat. The over $14 billion locked in several protocols is a statement of users' intentions.

On the other hand, just how fast NFT is growing, and its significance in blockchain is too hard to ignore.

Non-Fungible Tokens (NFTs) is--and will be a force to reckon with in 2021.

These tokens comply with the ERC-721 standard. NFT tokens are unique and can't be transferred. They are also not divisible.

Back to $777,777, and it is quite remarkable that one investor, Tim Kang, placed such a humongous bid, out-bidding everyone else after a back and forth, eventually having a pass to own a unique collection of digital art.

Tim was making a last-ditch effort.

He didn't want to miss out on the "Picasso" of digital art, so with just one second left, he made a bid of $777,777 securing ownership of "The Complete MF Collection."

It is made up of 20 pieces inspired by technology, Star Wars, and nature. The cool thing about this auction was that every piece of this collection was backed by a signed physical. Including a hair sample in each one to verify authenticity.

During the *Nifty Gateway*, Beeple-- real name Mike Winkelmann, went on to raise $3.5 million from the auction.

A big paycheck, yes. After all, "he had so much fucking fun" since he started working on his pieces 18 years ago.

Thing is, this won't be the last either.

Yes, we have had *Forever Rose* that went for $1 million, and other jaw-dropping NFT purchases.

According to Tommy Kimmelman, who heads the artist relation at *Nifty Gateway*, the NFT space, especially in Ethereum, has seen such a meteoritic rise in the past few years that such a sale was "inevitable" and that it won't be the last.

Kang shelling out $777,777 for a piece of digital art secured by cryptography is an endorsement of the space, highlighting its potential.

NFTs are largely unexplored. However, that doesn't mean they lack utility. Think about cryptocurrencies like Bitcoin earlier on. Only a few people knew what it was.

Combined with limited research into NFT awareness, how they can be used, and the unwillingness of people to experiment, it took eight years before the explosion of 2017.

Admittedly, NFT is still in a "discovery" space; many in the industry now know there is more to Ethereum and crypto than just moving money or raising funds. Tokenization and development of complex products are only one of them. Gaming, NFT, and other subsets will only mature with time.

Showing how much headroom there is for NFTs is the current trading volume of all projects to date.

It is sub-$150 million!!

From inception, just $147 million in trading volumes have been generated from the sale of slightly over 5.1 million pieces at an average price of $23, meaning there is so much more that NFTs have to offer.

Of all NFT platforms, *Super Rare* is popular. Through the platform, 259 pieces have been sold in the last week generating trading volumes of $474,656 paid directly to artists with security, provenance, and ownership visible straight on the blockchain.

There are no intermediaries or wait time or extra fees charged to artists; a divergence away from traditional systems where galleries determined which pieces made the cut.

From *Super Rare,* Micah Johnson, a former professional baseball player, sold a piece, sä-v(ə-)rən-tē, for $120k in December.

Overly, NFT is booming, and a foundation for future flights is being laid. Gradually but surely.

Month-to-month trading volumes are on the rise according to *DappRadar.*

It may take a few years before NFT dominate headlines and even outperform cryptocurrencies—the biggest application of blockchain technology. Banking on their simplicity, adaptability, and flexibility, NFTs now find utility in games—where they are used as in-game items, or even in DeFi—think of Rarible and even Yield Insurance, and more.

All this is because NFT has a deeper foundation and developing infrastructure following the rise of Cryptokitties in January 2018. Coupled with developments in Ethereum to reduce network fees, NFT will most likely find success, and therefore adoption---and more high-value auctions!

EOS - ETH: Which Suits You Better?

44

Written by Marco Gonzalez

Broadly speaking, EOS and Ethereum share philosophical similarities. ETH's early mover status, while certainly advantageous, is, in some aspects, a bit of a hindrance. When compared side-by-side with the younger, nextgen technology of EOS, ETH faults tend to amplify.

Make no mistake, ETH is powerful. Even the argument that it's more powerful than Bitcoin isn't entirely without basis. ETH's widespread smart contracts put it squarely in the public eye. For large corporations, ETH amassed somewhat of a reputation as a practical, standalone blockchain solution. The number of organizations exploring or actively leveraging Hyperledger tools is indeed significant.

Impending Upgrades

There's a lot on the horizon for the crypto industry. Where recent years saw an array of new projects and interesting perspectives, *2021 might be the year of the big chains. Bitcoin's Lightning Network* could shift more of the field it's the way. *ETH 2.0*, if deployed in a timely fashion, could go as far as propelling maximalists' visions. For the ETH community, much may rest on the day when practical operations are made available.

In terms of EOS's pragmatic approach towards blockchain solutions, there's robustness that defies critics. When the New Resource Model (Power Up) for the EOS public network is combined with EOSIO 2.1, market influencers likely re-examine their perspective. Factor in the *PRACTICAL* deployment time frames for:

- LN (BTC)
- ETH 2.0
- EOSIO 2.1+NRM

and network innovations appear to heavily favor EOS.

User Perspectives

The remainder of this examination focuses on user perspectives. Up until now, influences from developers and organizations shaped the markets. For example, *payment channels* are a driving force behind the *LN*. Critics of BTC's LN believe a hidden objective exists to support large retail operations. This is thought to be a detriment to the progress BTC has made over the years.

EOS and ETH communities seek to maintain diversity for new project development. Their impending upgrades provide balance and create an air of *symbiosis across networks.*

Moving toward global economic change has been forecasted by blockchain visionaries. Many believe it to be very real. A key factor will be users, especially onboarding those accustomed to fiat applications. How easy a fit a particular blockchain makes applications for daily life could prove decisive. For the time being, blockchain adoption remains in a sort of *pioneer stage.*

Factors to take into account:

- *Present* versus *future* applications
- Diversity of needs among users
- interrelated aspects

Present Applications
When it comes to present-day applications, EOS and Ethereum have each found a dependable level of viability.

Future Applications
Much of the future depends upon predictable delivery, and initial reception, of the new EOS and Ethereum versions.

Operational Costs
Surely EOS considered operational costs more than Ethereum during its creation. Some might even say that *ETH 2.0 is playing catch up.* In contrast, EOS 2.1 would fine-tune a proven performance engine. This isn't to say that EOS has been flying free

of turbulence. Just that ETH applications have not been nearly as simple to use as those powered by EOS.

Managing Accounts

From a user perspective, EOS maintains an agenda of *friendly* account management. Users complete transactions by simply remembering an account name, not a long chain of numbers. Additionally, a quick overview of the new EOS proposals reveals a wish to continue to refine management for both developers and users alike. Ethereum seems less concerned with streamlining blockchain managerial aspects.

Related Networks and Interoperability

Bridges have already been created across both ETH and EOS. However, interoperability slows when fees add up. Is it any wonder why ETH 2.0 aims to create a duality where fees become much less significant for *targeted* operations? Here again, is a position that EOS maintained from day one. When *deciding between the two*, keep in mind that it's unlikely that developers will ever choose to offload resources onto ETH as done by some already onto EOS

Some Clarification

For those that harbor even some Ethereum maximalist inclinations, please hold back any distaste just yet. Recall the early comparison of ETH 2.0 to Bitcoin's Lightning Network. The parallels may not initially seem straightforward. While ETH 2.0 is probably not going to draw many comparisons to the LN when it fully rolls out, *organizations running established payment channels may seriously consider both.* Even with this in common, however, Ethereum's vision still appears unique to that of the LN- or EOS for that matter.

Ethereum's first-mover advantage allows it to boast a network that competes for US dollar-based business applications, unlike any other blockchains. Distinguished, secure smart contracts and perceived market power have some believing that ETH may even rattle BTC dominance one day. Ironically, such a brash outlook could prove to limit if more nimble, practical smart contract applications rapidly grow in popularity with mainstream users. Such an outcome could alter fundamental market dynamics and user expectations.

There have been many Bitcoin-like philosophies adopted by new blockchains. Outside of major forks, new *core* blockchain innovation pales in comparison to that of smart contracts. EOS is certainly a powerhouse smart chain. Even discounting the many new projects leveraging either EOSIO or ETH technology, smart contract development is overall impressive.

Concluding Thoughts

Key issues users may wish to focus on when determining between EOS and ETH is the time frame and assets on hand. For comparatively robust applications *now*, EOS is a clear choice. Account management and operational costs win out over Ethereum today. EOS maintains these advantages for small or occasional transactions as well.

EOS updates are expected to impact the community within the coming weeks and months. Dependent projects and networks should notice stability never before seen in blockchain. For well-funded *Ethereum users*, security and first-mover networking seem to be the draw, especially since EOS can be used to offload less-critical transactions.

Lastly, keep in mind that there's a tendency to want to hold multiple cryptocurrencies and venture across chains out into the growing galaxy of crypto applications. Maximalist views are loud but remain a minority. *A network of blockchains is generally thought to be the future of the industry.* Maybe this is why EOS, as a third-generation blockchain, sought to ingrain account management at the heart of its development.

Intellectual Property and Publishing Rights For NFT's!

45

Written by Lars Kommienezuspadt

It was 1992 and seven renegade artists left Marvel to form Image Comics. Todd McFarlane, Jim Lee, Rob Liefeld, Whilce Portacio, Jim Valentino, Mark Silvestri, and Erik Larsen. It was a seismic event that shook the entire comic book industry. I still remember buying my first copies of Young Blood, Spawn, and The Darkness at my local comic store in Kingston, PA. The formation of Image Comics was a response to decades of unethical, exploitative practices by the publishers. Siegel and Shuster, witnessing millions of dollars being made from their creation, Superman, and the tragedy of Jack Kirby's exclusion to the rights of dozens of characters he created and co-created at Marvel. McFarlane, Liefeld, and the Image founders saw these inequities as cautionary tales, and thus, they responded, at the height of their popularity.

As a comic book artist myself, recounting the ugly history that is the relationship between creators and publishers, I do understand it. Artists, myself included, are inherently naive. We tend to ask the wrong questions, like, "Who owns this character?". Oftentimes, we're so excited just to get paid to draw, the questions we ask are, "Can I give him two plasma cannons?", or "What if she has a superpower she can't yet control?". Work for hire means just that. You are hired to create something and once it is produced, you're paid. End of transaction. Without predetermined agreements on copyrights and intellectual property considerations, the potential for creators to get written out of future profits is exponentially high.

This brings up our current climate as the NFT revolution is taking shape. It's the Wild West right now. Prior to writing this post, my friend Chris tweeted that Beeble, a well known digital artist, sold $500,000 worth of NFT's in less than two days! That's aspirational to almost all of us, but even on the NFT packs frontier like WAX, brands like Topps and Disney are raking in six figures per product launch. There is money to be made, for sure. This is just the latest in an endless string of blockchain trends, but NFT's are driven by creators. How can we ensure fairness for the artists? Could an Image Comics model be applied to the platforms curating and selling tokenized digital art?

Lately, I've been considering these questions after a recent exchange with my friend Julia Ponsford. Julia and I met on Steemit, and both of us invested time and effort into Ono [Oof! What a mess that was]. She's now heading operations at NFTShowroom.com, a fantastic tokenization platform built on Hive. A few months ago, I did a piece of promotional art for Dark County, a collectible, tradable gothic card game on WAX. EOS Detroit airdropped three variations of the art to Telegram users. It went over well and was widely received as EOSIO community goodwill. As a regular Hive contributor, I released 10 editions of the inked version on NFTShowroom and was subsequently recommended to burn them as they were already minted on a different platform.

Did you know that Marvel and DC would regularly destroy entire closets full of original comic pages from the '60s, '70s, and '80s? I unapologetically side with artists in these discussions, so I'm biased, but it seems that the publishers didn't need them after the plates were made, but didn't want the men and women that drew, inked, or colored the originals to profit from them either. That's changed. Even as recently as 2014, the courts ruled in favor of Jack Kirby's estate regarding a case brought by him in 1978. Sadly, he passed in 1994. Today, the typical arrangement includes companies having exclusive publishing rights, but artists retaining ownership rights which includes selling the original work, like you see here, listing full pages on Greg Capullo's official website, and prints; notably, non-sequential art. Pinup art, usually to be signed at conventions, etc.

My experience with NFTShowroom, I feel, falls under those existing norms within the comic book industry. I did artwork that was used by a company, published for promotional reasons. Yes, it was released on AtomicHub, but not by me as the artist. The promise of non-fungible tokens is the proof of an immutable origin, mint, and history, so from a collectors perspective, if two accounts mint their own versions of an NFT; one from a company that hired the artist, and one from the artist themselves, I would imagine the artist's token issuance would be the more collectible of the two. It's uncharted territory, but this is a conversation that needs to happen now. There's an urgent need to define what terms such as *publishing, ownership, exclusivity* mean in the NFT space.

Many of the articles here on Cryptowriter bring readers deep insights, data, and fact-based information on blockchain current events. This piece was meant to be part of a broader conversation. As I mentioned, I have a clear bias, so I welcome opposing points of view and I'd love to hear those in the comments section. Much like the comic book industry, the early business was messy. It took decades for disenfranchised creators to be recognized in the courts. My hope is, seeing that this is a developing space, we can avoid some of the quagmires that have plagued other industries, preserving mutual respect between artist and publisher relationships.

Has Bitcoin Found the Sweet Spot for Guaranteed Retail Adoption?

46

Written by Jason Deane

Back in the 1990s, I was directly involved with driving the adoption of the internet. Since then, the internet has developed into something very different from how we imagined it. It's also considerably better than we thought it would be.

While we dreamed of the giddy possibilities of a face to face video phone call or downloading a piece of software on demand, we never *ever* imagined web 3.0 apps like Uber or Airbnb. They were so far off the scale, they didn't even register.

Bitcoin's journey (and ours with it) is likely to be similar in the end, and some of the steps on that journey may not be linear, or even obvious.

Like most people, I found the fact that you could use Bitcoin to send money across any borders at any time without involving any banks or other organizations in just a few minutes (or worst case, hours) for a few dollars, simply astounding. This has never been possible before at any point in human history.

And it's not only a question of convenience and "people power" because as many as 1.7 billion people have no banking facilities at all (*source: World Bank Group Report 2020*) and Bitcoin—genuinely—has the power to fix this imbalance for good. What an amazing thought. It's one I still find genuinely exciting.

But in developed countries where we are spoiled for choice in terms of financial services, we hear people moaning about the fees of sending bitcoin all the time and suggestions of so-called new and better alternatives such as *[insert name of your favorite cryptocurrency which is "obviously" better than Bitcoin here].*

As new side and off-chain payment systems like the Lightning Network continue their development, I, like many others, considered this an essential factor for Bitcoin's widespread use going forward.

"When the Lighting Network goes fully live, it will be much better, and global adoption will happen," we say.

But, in light of how much the world and attitudes to the current financial system have changed—and so quickly—is that actually still the case?

Bitcoin is More Unique Than It First Appears

Long have we argued whether Bitcoin is a store of value or a currency and both points of view are technically correct in my view.

It is both. But it is also more. *Much* more.

Bitcoin is not only these things, it is also its own payment system.

It is the first time in human history that these important aspects have been combined and it puts us in entirely uncharted territory. It is inevitable that we will shape both our thinking and application of it as we continue to build and learn. Add in extreme scarcity and removing our imperfect human selves from the system completely, and we have something that will inevitably adapt in use as we do.

As the actions of institutions—including insurance companies, funds, and publicly listed corporates make the case against store of value increasingly difficult to argue, the next logical step will be for nation-states to treat Bitcoin in the same way.

Will it be a rogue nation such as Venezuela and Iran (both of which have been making noises along those lines recently) or perhaps a neutral, finance heavy economy such as Switzerland? It now seems more likely that it is a question of "who will be first" rather than "will it happen?"

And yet, since the same asset can also be used just as easily as a form of currency, does that de-value that purpose or reinforce it? Could it still be both? Should it be? Does it matter? What does that even look like?

Fiat—Bitcoin's New Friend

Bitcoin's famous scalability problem doesn't apply when used as a store of value on any institutional level and is, in fact, entirely irrelevant. It is a thousand times more efficient to move than gold—the classic treasury asset—and almost infinitely cheaper and quicker. Bitcoin works perfectly in this application and, in fact, it would be hard to design anything better if we tried to do it again today.

The issue has always been about retail spending, or what I call the "cold coffee" problem. That is, if you go into a coffee shop and buy a nice latte, it'll be cold by the time your transaction confirms and you can take it away. But if Bitcoin's own payment rail can't work well enough in that context, what do we do?

It's now possible that the solution may ultimately not be what we expected. This year has seen a blurring of the lines between our old financial system and the new crypto-based one, creating a working integration model that far exceeded any original expectations we may have had even as recently as 2019.

Whereas solutions designed to make Bitcoin usable in its native format such as BitPay and the Lightning Network seemed to be the way forward at one time, they have suddenly become less important.

For example, in the case of merchants receiving bitcoin for a transaction, it's easier for them to not have to work out how to deal with it at the point of sale and instead use a real-time fiat conversion system that sits inside standard accounting practices.

For the customer, they still get to use their bitcoin to make that transaction and the equivalent amount is deducted from their wallet using crypto debit cards or their chosen online payment system.

At the same time, the existing global payment rails, such as Visa, Mastercard, and PayPal, get to take a cut for providing their services. Even better, these are systems that most consumers already use every day and do so with confidence.

Not only does this solution allow the fiat-based and cryptosystems to seamlessly co-exist, they can even complement each other and thrive. It's also irrelevant to Bitcoin if fiat devalues as much as economists suspect it might, since a liquid, but the truly scarce asset will almost always provide a real-time "true" price on conversion.

The fact that you can buy and sell services in your local currency but at the level now set by the market in terms of exchange is both an astonishing and liberating thought, something that just wasn't possible a few years ago.

You could, for example, buy bitcoin through PayPal, let it sit in your account, and just spend little bits as it appreciates in value. Well, assuming the supply and demand curves play out that way of course.

And you can do this *right now*.

Passing Phase or Final Destination?

The purists won't like it, the libertarians will feel slightly cheated and the maximalists won't accept it, but I firmly believe this is a good thing.

It allows Bitcoin's purchasing power to normalize and become part of everyday transactions through familiar channels, while, at the same time, the store of value argument is being proven in the background.

These systems really can—and should—co-exist and it *still* levels the playing field across the globe. After all, any individual can buy and hold Bitcoin at any point, send it to anyone, and use its value to their advantage back in the traditional financial world with the real-time conversion. What's not to like?

The reality is that the growth in debit card services and the introduction of the PayPal bitcoin option mean that this is already happening. In fact, the use of those services is growing every day and may well now exceed the speed of growth of Lightning adoption, thereby winning the race for effective day-to-day use (if there even is one) in the meantime.

It could even be that Lightning ends up being no more than a faster way to do exactly what we did before—send money across borders without interference from third parties—rather than being a primary payment rail in, for example, a retail setting.

But who knows? It'll continue to evolve as we integrate the old system and the new, a process that is now all but unstoppable but is yet unknown in its final form.

Perhaps this is just a point we pass on the way to a Bitcoin standard, or perhaps it's as far as we go and this is where the Bitcoin train stops for good.

But wherever we go from here, Bitcoin's new, more direct relationship with fiat is almost certainly here to stay, at least for now.

And for end-users, that can only be a good thing.

Cryptocurrency & Generation Z - A Millennial Perspective on Financial Evolution

47

Written by Thomas Wolf

I woke up and said hello to my 18-year old nephew this morning, and while he was dealing with some phone issues, I asked him if he wanted to know anything about cryptocurrencies, what would it be?

"History is made by active, determined minorities, not by the majority, which seldom has a clear and consistent idea of what it really wants." - Ted Kaczynski

The same question most of us in the crypto world is slammed with routinely arose again as he asked "How do I get started?" It's not an easy answer, especially when I am NOT a financial advisor. I could answer the question in a hundred different ways and still feel I fell short, so instead, I am formulating a solution to why I cannot answer the question easily. I've told him to download Brave and starting gathering BAT; I've told him to get signed up on Coinbase and go through KYC verification and begin accepting Earn funds (Coinbase Earn now requires webcam/mobile pictures at the time of KYC, which I've now had to do twice) and to focus on the established currencies as many are near-worthless or going nowhere anytime soon.

Realizing this isn't an issue exclusive to my nephew, I've realized I have heard that question from all age groups. I began to explain that there are countless cryptocurrencies to influence his decision-making positively. Each has different functions and risks. It would take time to teach him each one, and more are being developed and expanded upon every day. I asked for a more specific question, but he had trouble forming one because he is entirely

unfamiliar with cryptocurrencies. In this, I see the real issue - he always has questions for me, so to see him struggle to find one tells me too much.

Generation Z represents the generation born after Millenials, also known as Centennials, born between 1996 and 2010. My nephew turned 18 a couple of months ago and should be precisely the type of demographic exchanges and other crypto services should be chasing. Instead, most of that money is going to millennials and even as far back as the Baby Boomers & Generation X - my parent's generation from back around 1965 and the generation before that.

Millennials are often targeted as "a trouble generation" in social media, and what's worse - I fear for them more than my generation in a variety of ways. I laugh about being considered a part of the problem by previous generations as they were the ones who handed my generation this mess.

A younger friend of mine from Vermont, belonging to this Generation Z demographic, shied away from becoming a writer in the Cryptowriter community despite being intelligent and interested. He was too concerned he was "too late into the crypto game." Despite my encouragement, he declined to humor the idea and thanked me for offering him to vouch for him. He felt he had already missed all the action and would only be wasting his time.

First of all, while those who held Bitcoin and other coins/tokens in their early days made out like bandits and became millionaires, the profits were only getting started. For whatever reason, many people in the younger generation seem to think that, because they weren't born during the "Bitcoin Boom" they have missed out on cryptocurrency-related opportunities. They could not be more mistaken as Bitcoin demonstrated us by breaching its all-time high recently and still holding at $19,145 USD not far below it, but that does not account for the sheer volume of money being held in a vast array of cryptocurrencies such as Ethereum (ETH), XRP (Ripple), Uniswap (UNI), and many other altcoins. There is more money in crypto now than there ever has been, and counting.

There is then the fact of cryptocurrency going mainstream with PayPal, Google, and many other corporations and merchants, but they are not quite there yet. The ongoing implementation process means Generation Z and generations after will be the generations we hand the keys to the castle to and educate them so they may continue to implement the currencies and blockchain tech we have worked so hard to develop and promote. The best way to educate the new generation is to point out flaws in our current financial system and quickly relate to the material rather than compare it to Robinhood stocks or CashApp, which takes them backward in time and progress.

Once their generation understands the problems we are attempting to solve with our previous generation's aftermath, they will see this is an evolutionary process. It is crucial for them to become a part of it for it to succeed. Crypto is a gift to the world; we can only keep it if we continue to give it away; without social support from the youth, our days would be quite numbered. Furthermore, people in this age bracket are crucial to keeping the progress we have made in forward momentum.

Not long ago, I wrote in response to Mark Helfman's article focusing on Bitcoin and Millennials, and shortly after, actions taken by the IRS reaffirmed my projections on how gray markets such as cryptocurrency would be handled by the US judicial system. The IRS is now putting cryptocurrency gains at the top of their 1040 tax forms, and my fears are realized: we need far more social support to move to true decentralization and move away from FIAT. Who will get us there? You? Me? All we've managed to do is keep cryptocurrencies popular and growing in use, and we are far from throwing our FIAT, aka "Legal Tender," in the fire for tinder, which is currently illegal.

No, real change doesn't happen so quickly on such a large scale too often. It will be Generation Z and those after that who make the decisions and the impact needed for financial freedom and equality. They must choose between two systems and decide for themselves how they want to operate when the day comes for one to outlive the other. For the sake of all people, I hope they succeed in realizing we have come to this crossroads for a reason. We do not want to be controlled. We want control over ourselves - and so does Generation Z; they just don't fully understand the scope of things yet, but they are very interested in learning!

The youth needs to understand that we are not just seeking to make money; we are seeking financial freedom from oppressive government regulations. Do you feel your government has been earning its taxes for the last four years, or far longer? I surely do not, but I still stay within the law. However, as I've said before - the laws are changing. Who will change them 20 years from now?

What we teach the children of today will inevitably lead to the outcome of this battle, a war among ourselves for autonomy. It's a shame people such as Ted Kaczynski (the Unabomber) were subjected to MK Ultra LSD mind-control experiments (that's a fact, it is declassified) only to be turned loose feeling lost in a hopelessly broken system, voiceless and suffering to the point where he turned away from society & technology entirely and began seriously injuring and killing people in order to get his writing published on a mere continental level.

I recommend the Netflix series "*Manhunt - Unabomber*" as it shows there was more to the story than some crazy man in the woods who liked to blow stuff up; that is a tragic waste of intelligent human life. I suppose it never occurred to him that in numbers, he could have had a voice forever while remaining free and non-violent.

With deep consideration I've decided to dissect Ted's book (arguably a manifesto, although written in a doctoral thesis format) and disprove every point made against technology in his "*Industrial Society and its Future*" and demonstrate that we can put an end to technology being used for the worse, being the only species that actively destroys its own environment, and use technology to improve human lives without dumbing down children glued to screens all day long, removing them from a natural setting.

Again, I find myself trying to reach common ground between people with very different idealisms. If I can prove it beyond a shadow of a doubt, I believe far more people would follow me than would follow a man locked in a cage full of rage. Despite what he has done to others, I feel terrible for that man, he reveres nature, is mentally ill and alone in his thoughts and suffering, likely to this day outside fan mail from neo-Luddites. Ted was born in 1942, also known as The Silent Generation. The irony of this is not lost on me.

We can have the best of both worlds. But, we need the numbers for it to happen. Kaczynski is a mathematical genius, and rather than working on blockchain tech, he was building bombs alone in a cabin. Can you imagine the impact he could have if he had gone the other way, the progress he could have offered the world? We are a minority, but in time we will be the majority; may we succeed in using truth as our weapon over violence.

Trading - The Mental Game

48

Written by Chris Cooper

#1 Fear of Missing Out (AKA, FOMO)

One of the most important concepts a new trader needs to grasp is that the market offers endless opportunities. Truly endless. Missing a good trade doesn't mean your life is over, nor does it even mean your trading career is over… You don't need to catch every trade to be profitable! Work on your mindset to realize when this fear is creeping in, and eliminate it when you can. Find what mentally works for you to get you "out of the rut". Simply wait for those 'perfect' setups to present themselves. The setups that follow your trading plan exactly. Allow your trading plan to be your filter, and don't get distracted or influenced by other trader's opinions or your own self-doubt. Trust your analysis and learn from each missed opportunity, allowing the knowledge that another is just around the corner to ease your fears.

Revenge Trading

The desire to be right 100% of the time and the inability to accept losses can lead to revenge trading. But what exactly is revenge trading? Simply put self-sabotage. After a trader experiences account drawdown or loss, feelings of frustration can surface, leading to the execution of reckless trades that don't meet their specific setup criteria.

Really, *any* trade that doesn't meet your setup criteria is reckless and is essentially gambling, but we'll delve more into that later. Regardless, self-sabotage is a very slippery slope and can lead to wiping out an entire account & leave you with naught but the knowledge that it was entirely your fault, which is even worse! Remember, losses are a part of the game! Accept them, learn from them, and move on. And don't assume you'll ever get good enough or learn from them to the point where you'll stop having them (losses), because that is purely a fantasy and misconception. Don't continue to place trades in the hopes of making your money back...that is the gist of it. In the long-term, this will not be sustainable and will only lead to more frustration and continual setbacks. Be patient and wait for your particular setup - the one that most easily catches your eye - to surface once more, and proceed accordingly.

Misusing Leverage

Too many new traders start off on their journey using leverage the wrong way. Risking 5-10% on each trade is an easy way to blow your account. That means you can only handle 10 losses before you're 100% bust. Instead, you may want to consider limiting your losses to 1 or 2% of your entire account size before entering a position. It may feel small and insignificant, but controlling your exposure is of the utmost importance if you plan on staying in the game long-term. I encourage my students to see themselves as Professional Risk Managers. And although losses are inevitable, as mentioned in the paragraph above, capital preservation is key. Absolutely paramount, in fact. Just because you may have the opportunity to use high leverage, such as 300:1, doesn't mean you should, nor does it mean it will accelerate your path to freedom. Likely it will just lead to the destruction of your account in ways you didn't even know existed.

Changing Trading Plans or "Style Drift"

Flip-flopping back and forth between different trading plans will not provide the results you're looking for… again, patience is key. It's much easier to place blame for your drawdown or lack of consistency on the educational program you learned from or on the strategies you're trading, but 99% of the time the problem is YOU. To be a trader you have to walk the line between so many things... you have to have humility yet have confidence. You must be hungry (have a desire for it), but fight off greed. I could go on for a full page with how hard it is mentally to become (and sustain *being*) a successful trader. Literally, the universe is against you, which is why less than 3% of traders are full-time and living off their accounts. So, take a step back, review what you've learned, and see where there are areas for improvement. Don't be so quick to jump ship and adopt an entirely new trading plan, because I guarantee there will be a steep learning curve for each one you implement. Perhaps you were just 1 tweak away from profitability before you gave up! Who is to know without persistence.

There are hundreds of ways to be profitable, yet taking 10% of 10 different profitable strategies can also result in a strategy that ultimately fails. You might be taking the wrong 10%! So instead of starting from zero with a new trading plan… be patient, allow your strategies to play out, and review each trade you take. Use forward-testing (trading with your strategy as opportunities arrive on the live market), but also back-testing (using historical data to determine if your strategy will likely be profitable in the future). There are tools for back-testing baked into many charting software like TradingView. This will ultimately lead to consistency and sustainability in the long-term.

If you identify with or have experienced any of the four chief trading mistakes I discussed above, great! You now have a broader awareness and can move forward, eliminating unnecessary losses and redundant problems that are inhibiting you. Don't allow your emotions to creep in, and *always* treat trading as a business. Would you show up 2-3 hours late to your job? Then don't show up 2-3 hours late to the charts if your strategy is dependent on your starting at a certain time... At the end of the day, you are a Professional Risk Manager, so act like one!

Why We Will Never Run Out of Bitcoin

49

Written by Jason Deane

Wait ... What?

With only 21,000,000 bitcoin ever going to be available, the logical conclusion is that we will "run out" of bitcoin if, for example, 21 million people decide they want to own one.

In fact, the situation is even worse than that since it will take another 120 years to mine the last 3.5 million coins AND we've collectively lost somewhere between 3.5 and 4 million coins (*source: Chainalysis*), almost certainly for good. In our lifetime, therefore, the best we can hope for is around 16 million coins being available.

Maybe.

Because even that assumes the early adopters with large balances, Grayscale, Microstrategy, PayPal, and an ever-increasing list of high net worth individuals plan to keep at least part of their newly acquired bitcoin in circulation, especially as they are currently absorbing more than miners are producing.

But this is probably not going to be the case.

Supply vs Demand

Bitcoin is an extremely unusual asset—by design of course—and the more you understand about it, the more incredible it reveals itself to be.

Bitcoin is both a money AND payment system built into one. Bitcoin is backed by network power and perfect mathematically executed transactions, which is far more than our day-to-day fiat currency actually is.

But most importantly, it is the only asset on the planet that has no supply variation at all, regardless of the demand level.

We're in completely uncharted territory here because, in every other case, as soon as there is a big enough increase in price, individuals, companies, and even sovereign states will move Heaven and Earth to produce some of the assets in question and profit from it. Supply increases, as a result, demand is eventually satisfied, price falls and equilibrium is restored. That's simply how market forces work.

But with Bitcoin this is impossible. Whether the fiat equivalent of bitcoin is $1 or $1,000,000, the same exact number will be produced in each time period, currently 900 new coins a day, down from 1800 in May after the last halving. And there is not a single thing any individual, organization, or even country can do about it, no matter their power or influence.

This extraordinary situation has always had the potential to create problems with market dynamics.

Initially, this meant an excess supply because, in the early days, enormous amounts of bitcoin were produced (production started at 7,200 a day back in 2009) at a time where no-one wanted to take them. So desperate were the early communities to create demand, they were given away for free from 'faucets'—a fact that seems incredulous in these different times.

Now, however, we are entering a new era. Not only are we on the cusp of global "user" adoption, initially through limited (but user familiar) services such as PayPal, but also on the cusp of significant institutional adoption. As each day passes, new names join the list.

First, MicroStrategy, then Square, then a slew of other organizations with varying levels. Today, as I write this article, the Massachusetts Mutual Life Insurance Co announced its own investment into Bitcoin of $100m. That's an enormous "play it safe" insurance conglomerate investing in Bitcoin—unthinkable only six months ago.

But here's the problem:

These are very large purchases that are hoovering up bitcoin at a faster rate than it can be produced. Not only that, but many of these organizations have made it clear that they intend to keep it "for the long term" or to create a "Bitcoin standard." This means that anything they acquire is not likely to make it back to the market any time soon.

Logically speaking, we must therefore run out of bitcoin. And probably quite soon.

But is that really the case?

"Infinite" Bitcoin

Here's where it gets interesting. Although Bitcoin's *total* supply is strictly limited, the *market* supply actually isn't.

There will always be bitcoin available to purchase should you wish to do so. But with no increase in production possible, it's just a question of what you're prepared to pay.

In short, bitcoin availability is determined entirely by demand. That demand sets the price and in turn, that price sets the level of supply.

At the moment, for example, the collective purchases of PayPal, Grayscale, MicroStrategy, and others are exceeding the amount of new bitcoin that is entering the market each day. The result is a demand-driven price increase to the levels we have seen thus far. As more players enter the market, the upwards pressure is only likely to increase.

Since extra bitcoin is required to fulfill those orders that are over and above what the miners produce, existing holders must be tempted to part with theirs. They are tempted, quite simply, by higher fiat exchange rates.

The price at which each seller is prepared to release their bitcoin is entirely dependent on what their agenda is. There are, for example, many early adopters who hold vast amounts who have, as yet, not been tempted to release it to the market. However, as the old adage goes:

All men have their price and all women their figure

At some point, it is likely that some of that bitcoin will be released when the fiat equivalent is high enough. And, of course, the higher that price is, the less of their reserve they need to achieve substantial wealth in fiat terms.

The reality is that the days of easy and relatively cheap bitcoin acquisition are probably coming to an end.

The "Everyone HODLs" Scenario

There is one other aspect to this. Since Bitcoin is so scarce, it is theoretically possible (although unlikely given the market forces above) that almost all bitcoin could be put into cold storage and never recirculated.

But even if just a few Satoshi were left in the open market, it would become the new equivalent of 'Bitcoin.' Effectively, since Bitcoin is entirely digital, we would simply create new decimals of the very same asset we have before, dividing it infinitely, and repricing each unit as the market dictates.

At the moment, quantities of bitcoin are usually represented in an eight-digit format thus:

0.00000001

but in this scenario, we'd simply extend it to any level that the market considers appropriate. In theory, it could be any limit and we'll probably even "coin" (pun intended) a new name for the decimal points of a Satoshi. Perhaps we should start taking suggestions now.

Of course, at no time would any additional bitcoin actually be produced, we'd just be moving the decimal point and repricing it accordingly, thereby perfectly demonstrating the power and flexibility of digital scarcity.

In the physical world, this is just not possible. You could, for example, keep cutting gold bars in half and repricing them in the same way, but there is a point where this becomes a practical impossibility. Bitcoin is simply a far more elegant and efficient solution.

In reality, the chances are that it will simply be too expensive and too valuable for the majority of people to use or own entire bitcoin in the future should the current rate of adoption continue as is. Instead, most will use whatever fraction becomes the norm.

But at the end of the day, does that even matter?

After all, whatever unit we actually use, the strength, power, and certainty of Bitcoin will transcend the decimal point entirely.

And that brings reassurance at any level.

The Biggest Danger for the Coming Years Is Not Owning Bitcoin

50

Written by Sylvain Saurel

The year 2020 will forever be remembered as a special year for the world. In a few weeks, all the certainties we had were gone. The coronavirus pandemic has clearly reshuffled the cards.

On a scale not seen in decades, this pandemic showed how unprepared we were for such an event. The governments of the world's major economic powers had not anticipated anything. Yet it is the role of our politicians to prepare for such worst-case scenarios.

But again, they failed.

Many now believe that pandemics of this type are the greatest dangers we will face in the years to come. Indeed, it is certain that other viruses such as COVID-19 will spread around the world at great speed in the future.

Others will tell you that the biggest danger for the coming years will be the cold war between the United States and China for the status of the world's leading superpower.

The United States is currently, by far, the world's leading economic and military power. Nevertheless, China's current trajectory suggests that the country of Xi Jinping will be able to overtake the United States as early as 2030. In any case, this is the objective pursued by Xi Jinping.

To achieve this, Xi Jinping is carrying out various ultra-ambitious projects, with the Belt and Road Initiative, but also the e-RMB. China is ahead of all other countries in the world in the field of central bank digital currencies (CBDC). With its e-RMB, which is already in full-scale testing in major Chinese cities, China aims to break the monopoly of the U.S. dollar.

There is no guarantee that China will succeed. However, we can recognize that China is giving itself the means to achieve its ambitions.

The Cold War between the United States and China is therefore inevitable. Some even fear that this war will become a reality sooner or later. This is indeed a possible scenario. It is more pessimistic, I grant you, but many fear it.

Other people have much more superficial problems in their lives. These people will tell you that the biggest danger they perceive is, for example, having their social media account banned. For some people, losing their Twitter, Instagram, or Facebook account would be a tragedy.

I hope that people who feel this way are in the minority, but when I see how some people are addicted to their smartphones and social networks, I have serious doubts.

Of all the dangers that people perceive for the coming years, the biggest in my opinion is not owning Bitcoin.

People who read this statement will think that I am foolish to say such things. These people do not realize the real dangers that await them in the future. Pandemics like coronavirus are one thing, the danger of a war between two great world powers is a worrisome prospect.

But the prospect of seeing your hard-earned wealth stolen away by monetary inflation is something even more worrisome. Worse still is the prospect of the mass surveillance society that many states may be tempted to establish in the future.

Once CBDC's are deployed all around the world, States will seek to eliminate cash so that they can better control your money. They will be able to confiscate what you have even more easily than they can now with the banking system. They will be able to censor your transactions for totally arbitrary reasons.

Some people think they are safe because they live in a great Western democracy. Unfortunately, these people seem to forget that freedom is never really acquired. One must always fight to keep it. This is your duty as a citizen.

What is currently happening in China with the establishment of a mass surveillance society may very well happen one day soon in European countries or the United States. If citizens do not fight to keep their freedom, politicians will soon be taking a little bit more from them each time.

To protect yourself, you must opt for Bitcoin. The electronic money invented by Satoshi Nakamoto is the best weapon at your disposal to protect your wealth in the face of the great monetary inflation in a censor-proof way.

By purchasing Bitcoin, and holding the private keys associated with your BTC, you will be in total control. You will protect your hard-earned money from both monetary inflation and censorship. You will be able to opt for a low time preference.

In a world of the future where everything will be digital, Bitcoin will be your weapon to ensure your freedom when it comes to money. Few people realize it yet, but it is something essential. Over time, more and more people will realize it.

Once you have joined the camp of those who have understood why buying Bitcoin is vital for the years to come, you will realize that my assertion was far from exaggerated. At that point, you will be ready to change your life forever by buying Bitcoin.

Made in the USA
Monee, IL
04 November 2026

67378314R00167